Lecture Notes
in Economics and
Mathematical Systems

Operations Research, Computer Science, Social Science

Edited by M. Beckmann, Providence, G. Goos, Karlsruhe, and
H. P. Künzi, Zürich

67

I. V. Girsanov

Lectures on Mathematical Theory of Extremum Problems

Springer-Verlag
Berlin · Heidelberg · New York 1972

Igor Vladimirovich Girsanov†

Edited by
Prof. B. T. Poljak
Moscow State University
Computer Center
Moscow V-234/USSR

Translated from the Russian by
D. Louvish
Israel Program for Scientific Translations
Kiryat Moshe
P. O. Box 7145
Jerusalem/Israel

AMS Subject Classifications (1970): 46N05, 49B30, 49B40, 52A40

ISBN 3-540-05857-5 Springer-Verlag Berlin Heidelberg New York
ISBN 0-387-05857-5 Springer-Verlag New York Heidelberg Berlin

Offsetdruck: Julius Beltz, Hemsbach/Bergstr.

Extremal problems are now playing an ever-increasing role in applications of mathematics. It has been discovered that, notwithstanding the great diversity of these problems, they can be attacked by a unified functional-analytic approach, first suggested by A. Ya. Dubovitskii and A. A. Milyutin. The book is devoted to an exposition of this approach and its application to the analysis of specific extremal problems. All requisite material from functional analysis is first presented, and a general scheme for derivation of optimum conditions is then described. Using this scheme, necessary extremum conditions are then derived for a series of problems — ranging from Pontryagin's maximum principle in optimal control theory to duality theorems in linear programming.

The book should be of interest not only to mathematicians, but also to those working in other fields involving optimization problems.

TABLE OF CONTENTS

EDITOR'S PREFACE

The author of this book, Igor' Vladimirovich Girsanov, was one of the first mathematicians to study general extremum problems and to realize the feasibility and desirability of a unified theory of extremal problems, based on a functional-analytic approach. He actively advocated this view, and his special course, given at the Faculty of Mechanics and Mathematics of the Moscow State University in 1963 and 1964, was apparently the first systematic exposition of a unified approach to the theory of extremal problems. This approach was based on the ideas of Dubovitskii and Milyutin [1].

The general theory of extremal problems has developed so intensely during the past few years that its basic concepts may now be considered finalized.

Nevertheless, as yet the basic results of this new field of mathematics have not been presented in a form accessible to a wide range of readers. (The profound paper of Dubovitskii and Milyutin [2] can hardly be recommended for a first study of the theory, since, in particular, it does not contain proofs of the fundamental theorems.)

Girsanov's book fills this gap. It contains a systematic exposition of the general principles underlying the derivation of necessary and sufficient conditions for an extremum, in a wide variety of problems. Numerous applications are given to specific extremal problems. The main material is preceded by an introductory section in which all prerequisites from functional analysis are presented.

Girsanov had long intended to write this monograph, but his tragic death in March 1967 prevented him from carrying out this project. The present book is based on Girsanov's lecture notes, which were published in lithograph form in 1964. Since these notes were not prepared by the author for the press, I took the liberty of making a few changes in the text and adding references.

B. T. Polyak

INTRODUCTION

Extremal problems were the object of mathematical research at the very earliest stages of the development of mathematics. The first results were then systematized and brought together under the heading of the calculus of variations, with its innumerable applications to physics and mechanics. Attention was devoted principally to the analysis of smooth functions and functionals defined over the entire space or restricted to some smooth manifold. The extremum conditions in this case are the Euler equations (with Lagrange multipliers in the case of constraints).

Independently of the calculus of variations, the theory of approximations was developed; the methods figuring in this theory, especially in the theory of Chebyshev approximations, had a specialized nature.

Technological progress presented the calculus of variations with a new type of problem — the control of objects whose control parameters are varied in some closed set with boundary. Quite varied problems of this type were investigated by Pontryagin, Boltyanskii, Gamkrelidze and Mishchenko, who established a necessary condition for an extremum — the so-called Pontryagin maximum principle. The nature of this condition and the form of the optimal solutions were so different from the classical theorems of the calculus of variations that popular-science writers began to speak of the advent of a "new" calculus of variations.

Something similar happened in the realm of extremal problems for functions of a finite number of variables. Economic necessity gave rise to the appearance of special methods for determining the extrema of smooth functions on closed domains with piecewise-smooth boundaries. First results in this direction were obtained in

1939 by Kantorovich. This field of mathematics is now known as mathematical (nonlinear) programming.

That the results and methods of mathematical programming were similar to the Pontryagin theory was obvious, but the subtle and elegant geometric technique of Boltyanskii and Pontryagin somewhat obscured the analytical content of the problem.

Finally, at the end of 1962, Dubovitskii and Milyutin found a necessary condition for an extremum, in the form of an equation set down in the language of functional analysis. They were able to derive, as special cases of this condition, almost all previously known necessary extremum conditions and thus to recover the lost theoretical unity of the calculus of variations.

The present lecture course will be devoted primarily to the study of extremal problems in the framework of the general Dubovitskii-Milyutin theory, though we shall also consider results due to other authors.

We now list some of the extremum problems to be studied during this course.

1. Let $F_0(x)$ be a smooth function in the space R^m, and $Q \subset R^m$ a set defined by a system of equations $F_i(x) = 0$, $i = 1, \dots, n$. Find necessary conditions for the function $F_0(x)$ to assume its minimum on Q at a point $x_0 \in Q$. This problem is solved in classical analysis by the introduction of Lagrange multipliers.

2. Now let the set Q be defined by a system of equations $F_i(x) = 0$, $i = 1, \dots, k$ and inequalities $F_i(x) \leqslant 0$, $i = k+1, \dots, n$. This yields the general problem of nonlinear programming. Necessary conditions for an extremum in this case were obtained only relatively recently. A special case is the problem of linear programming: find min (c, x) under the condition $Ax \leqslant b$ ($x, c \in R^m$, $b \in R^n$, A is an $n \times m$ matrix). In this case the necessary extremum conditions turn out to be sufficient as well.

3. Let $x(t)$ be a differentiable function, $\Phi(x, y, t)$ a smooth function of three

variables. Find a function x(t) such that the functional

$$\int_{t_0}^{t_1} \Phi\left(x(t), \frac{dx(t)}{dt}, t\right) dt$$

is a minimum, under the condition $x(t_0) = c$, $x(t_1) = d$. This is the fundamental problem of the calculus of variations; necessary conditions for an extremum are given by the Euler equation. Other conditions can be introduced instead of rigidly fixed endpoints.

4. Let $x_1(t)$, $x_2(t)$ be a family of curves in three-dimensional space, lying on a surface $G(x_1, x_2, t) = 0$.

Find a curve in this family which minimizes the integral

$$\int_{t_0}^{t_1} \Phi\left(x_1(t), x_2(t), \frac{dx_1(t)}{dt}, \frac{dx_2(t)}{dt}, t\right) dt$$

(the Lagrange problem). Here the answer is given by the Euler equation with additional terms to allow for the constraint $G = 0$.

5. Find the minimum of

$$\int_{t_0}^{t_1} \Phi(x(t), u(t), t)\, dt,$$

where $x \in R^n$, $u \in R^r$,

$$\frac{dx(t)}{dt} = \varphi(x(t), u(t), t),$$

$$x(t_0) = c, \; x(t_1) = d.$$

This is the so-called Lagrange problem with nonholonomic constraints. The solution is analogous to that of the preceding problem.

6. Under the assumptions of Problem 5, we impose an additional constraint: $u(t) \in M$, $t_0 \leqslant t \leqslant t_1$, where M is some set in R^r. This problem, known as the problem of optimal control, was considered by Pontryagin and his students; the necessary extremum conditions constitute the Pontryagin maximum principle.

7. Let $\varphi_i(t)$, $i = 1, \ldots, n$ be continuous functions on $[t_0, t_1]$. Given a

continuous function $\varphi_0(t)$, find a vector $x = (x_1, \ldots, x_n)$ such that the function

$$F(x) = \max_{t_0 \leqslant t \leqslant t_1} |\varphi_0(t) - \sum_{i=1}^{n} x_i \varphi_i(t)|$$

is a minimum. This problem is studied in the theory of best approximations. Its distinguishing feature is the fact that the function F(x) is not differentiable.

8. Let x(t) and u(t) be two functions such that

$$\frac{dx(t)}{dt} = \varphi(x(t), u(t), t)$$

and G(x, t) some continuous function. Determine $u(t) \in M$ such that $\max_{t_0 \leqslant t \leqslant t_1} G(x(t), t)$ is a minimum (in other words, find $\min_u \max_{t_0 \leqslant t \leqslant t_1} G(x(t), t)$). This problem also involves a nondifferentiable functional

$$F(u) = \max_{t_0 \leqslant t \leqslant t_1} G(x(t), t).$$

It turns out that all these problems can be studied in a unified, general scheme; the idea of this scheme will now be described.

Let E be a linear space and F(x) some functional on E. Let Q_i, $i = 1, \ldots, n$ be sets in E. Let $x \in Q = \bigcap_{i=1}^{n} Q_i$ be a point at which F(x) assumes a minimum on Q. In this very general situation, let us try to formulate necessary conditions for an extremum. Consider the set $Q_0 = \{x: F(x) < F(x_0)\}$. An obvious necessary (and sufficient) condition is then

$$\bigcap_{i=0}^{n} Q_i = \varnothing. \tag{1.1}$$

However, this statement is of no practical value for the investigation of extremal problems. In order to lend it more substance, one must impose restrictions on the classes of sets Q_i and functionals F(x) considered. Let us consider a few examples, with a view to choosing sufficiently natural restrictions.

We begin with the simplest case: find the minimum of $y = F(x)$, $x \in R^1$.

Suppose that the minimum is assumed at a point (x_0, y_0). Then $Q_0 = \{x, y: y < y_0\}$. If we denote the graph of $y = F(x)$ on the plane by Q_1, the condition $Q_0 \cap Q_1 = \emptyset$ expresses the trivial fact that no points of the graph $y = F(x)$ lie below the straight line $y = y_0$ (Fig. 1). Now assume that F(x) is a differentiable function. Set $\bar{x} = x - x_0$, $\bar{y} = y - y_0$, $z = (\bar{x}, \bar{y})$. Then the equation of the tangent l to $y = F(x)$ at the point (x_0, y_0) is $\bar{y} - F'(x_0)\bar{x} = 0$, and Q_0 is defined by the condition $-\bar{y} > 0$. If we define vectors $f_1 = (-F'(x_0), 1)$ and $f_0 = (0, -1)$, we can define l and Q_0 as follows: $l = \{z: (f_1, z) = 0\}$, $Q_0 = \{z: (f_0, z) > 0\}$. The usual necessary condition for an extremum, $F'(x_0) = 0$, is expressed in this setting as

$$f_0 + f_1 = 0.$$

Now consider a more complicated example. Find the condition for the minimum of a smooth function F(x), $x = (x_1, x_2)$, on the set $Q_1 = \{x: x_1 \geqslant 0, x_2 \geqslant 0\}$. Suppose that the minimum is assumed at the point $\overset{0}{x} = 0$. Let F'(x) denote the gradient of the function F(x):

$$F'(x) = \left(\frac{\partial F(x)}{\partial x_1}, \frac{\partial F(x)}{\partial x_2}\right).$$

To simplify matters, we first consider the case in which the function F(x) is concave. Then, if $F'(0) \neq 0$, the set $Q_0 = \{x: F(x) < F(0)\}$ will always contain the half-plane $K_0 = \{x: (F'(\overset{0}{x}), x) < 0\}$. Introducing the unit vectors $a_1 = (1, 0)$, $a_2 = (0, 1)$, we can express Q_1 as the solution of a system of inequalities: $Q_1 = \{x: (a_1, x) \geqslant 0, (a_2, x) \geqslant 0\}$. A necessary condition for the point $\overset{0}{x} = 0$ to be a minimum point for F(x) on Q_1, by (1.1), is that $Q_0 \cap Q_1 = \emptyset$, and a necessary condition for this, in turn, is that Q_1 be disjoint from K_0. It is easily seen (Fig. 2) that a necessary and sufficient condition for this to hold is that $F'(0) = \lambda_1 a_1 + \lambda_2 a_2$, where $\lambda_1 \geqslant 0$, $\lambda_2 \geqslant 0$. We now notice that, for nonnegative λ_1, λ_2, $(\lambda_1 a_1 + \lambda_2 a_2, x)$ is the general form of a linear function which is nonnegative on Q_1, and $\lambda_0(F'(0), x)$ is that of a linear function nonnegative on K_0. Setting $F'(0) = f_0$, $\lambda_1 a_1 + \lambda_2 a_2 = f_1$, we get the equation $f_0 + f_1 = 0$, where $(f_0, x) \geqslant 0$ on K_0, $(f_1, x) \geqslant 0$ on Q_1, and moreover f_0, $f_1 \neq 0$ in the case $F'(0) \neq 0$.

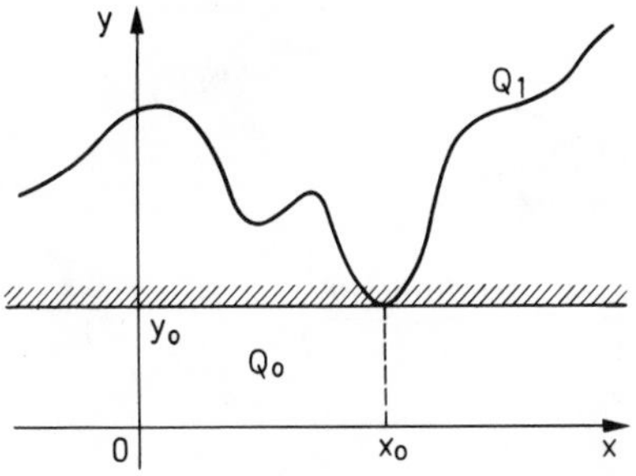

Fig. 1

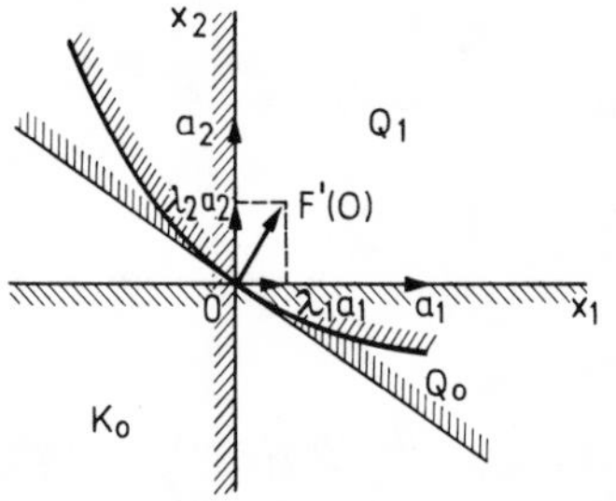

Fig. 2

Now this result can be interpreted in a different way. Let us regard the set $Q_1 = \{x: (a_1, x) \geqslant 0,\ (a_2, x) \geqslant 0\}$ as the intersection of the sets $K_1 = \{x: (a_1, x) \geqslant 0\}$ and $K_2 = \{x: (a_2, x) \geqslant 0\}$. It is readily seen that the functions $\lambda_1(a_1, x)$ and $\lambda_2(a_2, x)$ are linear, and nonnegative on K_1 and K_2, respectively. Setting $\lambda_1 a_1 = f_1$ and $\lambda_2 a_2 = f_2$, we get $f_0 + f_1 + f_2 = 0$, where $f_0 \geqslant 0$ on K_0, $f_1 \geqslant 0$ on K_1 and $f_2 \geqslant 0$ on K_2. If the reasoning is carried out a little more cautiously, it can be seen that these results are also valid when the function $F(x)$ is not concave.

To conclude this discussion, we consider an example which involves both equations and inequalities. Let the set Q be defined in R^3 as the intersection of the following three sets:

$$Q_1 = \{x: x_1 \geqslant 0\},\ Q_2 = \{x: x_2 \geqslant 0\},\ Q_3 = \{x: G(x) = 0\},$$

where $x = (x_1, x_2, x_3)$ and $G(x)$ is a differentiable function such that $G(0) = 0$.

Let $F(x)$ be a smooth function. Find necessary conditions for $F(x)$ to assume a minimum on Q at the point 0, i.e., for $Q_0 \cap Q = \emptyset$, where $Q_0 = \{x: F(x) < F(0)\}$. Assume that the equation $G(x) = G(x_1, x_2, x_3) = 0$ can be solved for x_3 in a neighborhood of the point $(0, 0, 0)$: $x_3 = \tilde{G}(x_1, x_2)$. Then our problem reduces to finding extremum conditions for the function $\tilde{F}(x_1, x_2) = F(x_1, x_2, \tilde{G}(x_1, x_2))$ on the set

$Q_1 \cap Q_2$ at the point (0, 0). We already know that a necessary condition here is

$$-\tilde{F}'(0) + \lambda_1 a_1 + \lambda_2 a_2 = 0;\ \lambda_1 \geqslant 0,\ \lambda_2 \geqslant 0;\ a_1 = (1,0),\ a_2 = (0,1).$$

But

$$\tilde{F}'(0) = \left(\frac{\partial \tilde{F}(0)}{\partial x_1}, \frac{\partial \tilde{F}(0)}{\partial x_2}\right) =$$

$$= \left(\frac{\partial F(0)}{\partial x_1} + \frac{\partial F(0)}{\partial x_3} \cdot \frac{\partial \tilde{G}(0)}{\partial x_1}, \frac{\partial F(0)}{\partial x_2} + \frac{\partial F(0)}{\partial x_3} \cdot \frac{\partial \tilde{G}(0)}{\partial x_2}\right),$$

while $\dfrac{\partial \tilde{G}(0)}{\partial x_i} = -\dfrac{\partial G(0)}{\partial x_i} \Big/ \dfrac{\partial G(0)}{\partial x_3}$, $i = 1,2$.

Thus, setting $\lambda_3 = \dfrac{\partial F(0)}{\partial x_3} / \dfrac{\partial G(0)}{\partial x_3}$, $f_0 = -F'(0)$, $f_1 = (\lambda_1, 0, 0)$, $f_2 = (0, \lambda_2, 0)$, $f_3 = \lambda_3 G'(0)$, we get the necessary extremum condition in its final form:

$$f_0 + f_1 + f_2 + f_3 = 0, \tag{1.2}$$

where $(f_i, x) \geqslant 0$ on Q_i, $i = 0, 1, 2$, and $(f_3, x) = 0$ on the subspace tangent to Q_3.

Thus, in all cases considered we have been able to derive necessary conditions for an extremum in the form of a certain equation for linear functions, which are related to the constraints by a nonnegativity condition. Moreover, the linear function related to the original functional figures in the final formula (1.2) in the same way as the other functions.

As a conclusive argument toward the thesis that condition (1.1) is more suitably expressed as an equation (1.2), we shall see when (1.2) implies (1.1). The following lemma is obvious:

Lemma 1.1. Let Q_i, $i = 0, \ldots, n$ be certain sets in E, (f_i, x) linear functions such that $(f_i, x) \geqslant 0$ on Q_i and, for at least one i, $(f_i, x) \neq 0$ on Q_i. Then the condition $f_0 + \ldots + f_n = 0$ implies that $\bigcap_{i=0}^{n} Q_i = \emptyset$.

The following question is natural in this context. For what sets and in what spaces can condition (1.1) be expressed in analytical form using linear functions which are nonnegative on Q_i? There is a simple and fairly extensive class of such sets — convex cones (see Lecture 5). It turns out that a system K_i, $i = 1, \dots, n$, of open convex cones with apex at 0 has empty intersection if and only if there exist linear functions $f_0, f_1, \dots, f_n$, not all zero, such that $f_i(x) \geqslant 0$ for $x \in K_i$ and

$$f_0 + f_i + \dots + f_n = 0. \tag{1.3}$$

This result can be sharpened: one of the cones need not be open — it can be any convex cone, in particular, a linear manifold. The equality (1.3) is almost self-evident for the case of two cones. Indeed, let $K_1 \cap K_2 = \emptyset$. It is then clear from geometric considerations (Fig. 3) that there exists a hyperplane H separating the cones and passing through their common apex. Let $f(x) = 0$ be the equation of this hyperplane. Then $f(x) \geqslant 0$ on the cone K_1, $f(x) \leqslant 0$ on the cone K_2, or vice versa. Setting $f_1(x) = f(x)$, $f_2(x) = -f(x)$, we get the desired result.

In the examples considered above, equality (1.3) has an obvious meaning. To derive meaningful analytical conclusions from the general relation (1.3), one needs an adequately complete description of the family of linear functions nonnegative on any given system of cones K_i.

In order to do this, and also in order to prove (1.3) under fairly general conditions, we shall have to make a lengthy digression into the field of functional analysis. This will occupy us in Lectures 2 through 5. In Lecture 6 we shall derive a general formulation of the necessary conditions for an extremum. The four following lectures (7 through 10) will be devoted to a description of the technique employed in applying these general conditions to concrete problems. Then, in Lectures 11 through 14, we shall derive necessary conditions for a minimum in the various problems listed above. Finally, in the concluding lectures (15 and 16) we shall show that in the case of convex problems the necessary extremum conditions are also sufficient, and present other formulations of the extremum conditions for these problems.

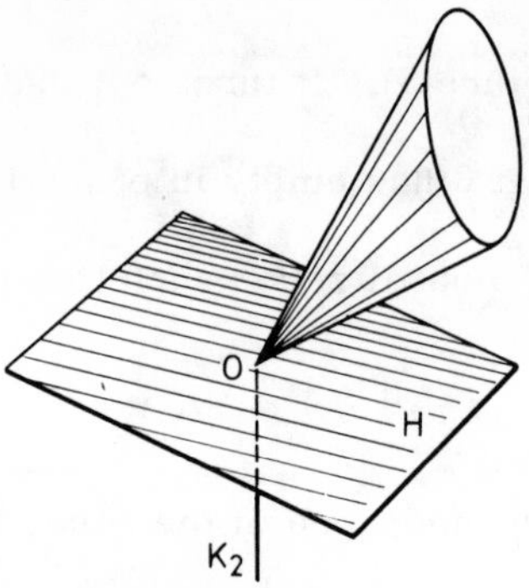

Fig. 3.

We should like to emphasize that the results presented below by no means exhaust the general theory of extremal problems. Among the parts of the theory omitted here are the problem of existence and uniqueness of the extremum, the stability of the extremum, the theory of the second variation, methods of numerical solution, and other problems.

TOPOLOGICAL LINEAR SPACES, CONVEX SETS, WEAK TOPOLOGIES

We shall assume that the space in which the extremal problem is to be solved is a topological linear space. We briefly recall the relevant concepts.

A set E of elements x is said to be a topological linear space if, first, it is a linear space (i. e., addition of elements and multiplication of elements by real numbers (scalars) are defined and satisfy the natural conditions: commutativity, distributivity, etc.); second, it is endowed with a topology; third, addition and multiplication by scalars are continuous functions in the given topology. Elements of E will generally be denoted by lower case roman letters, scalars by Greek letters, sets in E by roman capitals, families of sets by script capitals.

The topology of the space E is defined by specifying the family of open sets, which have the following properties: the empty set $\emptyset$ and the whole space E are open sets; the union of any number of open sets and the intersection of finitely many open sets are open sets. The complement of an open set is a closed set. Any open set containing a point x is called a neighborhood of x. In order to define a topology it is sufficient to specify a basis $\mathcal{A}(x)$ about each point x, i. e., a family of neighborhoods of x such that, for any neighborhood V of x, there exists $U \in \mathcal{A}(x)$ such that $U \subset V$. With this definition, a set Q is open if and only if for every $x \in Q$ there exists $U \in \mathcal{A}(x)$ such that $U \subset Q$. In a linear space, it is sufficient to specify a basis $\mathcal{A}$ about zero, for then the family $x + \mathcal{A}$ is a basis about x. With this definition, every $U \in \mathcal{A}$ is an absorbing set, i. e., for any $x \in E$ there exists $\lambda > 0$ such that $\lambda x \in U$. Moreover, one may always assume that the basis about zero is symmetric, i. e., if $U \in \mathcal{A}$, $x \in U$, then $-x \in U$. If x is a point of a set Q and there exists

$U \in \mathcal{A} + x$ such that $U \subset Q$, then x is called an interior point of Q. The set of all interior points of Q is denoted by Q^0. The closure $\overline{Q}$ of a set Q is the set of all points x such that $U \cap Q \neq \emptyset$ for all $U \in \mathcal{A}(x)$. If $x \in \overline{Q}$ but $x \notin Q^0$, then x is called a boundary point of Q. Obviously, $Q = Q^0$ if and only if Q is open, $Q = \overline{Q}$ if and only if Q is closed. Recall also that a set Q is said to be bounded if, for every $U \in \mathcal{A}$, there exists λ such that $Q \subset \lambda U$. As usual, the notation $A + B$, $A - B$, λA for sets in a linear space has the following meaning:

$$A+B=\{x+y,\ x\in A,\ y\in B\},\ A-B=\{x-y,\ x\in A,\ y\in B\},$$
$$\lambda A=\{\lambda x,\ x\in A\}.$$

In general, $A + A \neq 2A$, $A - A \neq 0$ [note that here 0 denotes not the empty set $\emptyset$ but the set consisting of the single element 0], and the sum of two closed sets need not be closed.

Let E and E_1 be topological spaces. An operator P(x) mapping E into E_1 is said to be continuous at a point $x \in E$ if, for any neighborhood V of P(x), there exists a neighborhood U of x such that $P(U) \subset V$. If P(x) is continuous at all points of E, P(x) is said to be continuous (on E). P(x) is continuous if and only if the preimage of every open (closed) set is open (closed). If E_1 is the real line (i.e., the values of P(x) are real numbers), we call P(x) a functional. We generally use the notation F(x), G(x), etc. for functionals. A functional F(x) is continuous if and only if all the sets $\{x\colon F(x) < \lambda\}$ and $\{x\colon F(x) > \lambda\}$ are open (and all the sets $\{x\colon F(x) \leq \lambda\}$ and $\{x\colon F(x) \geq \lambda\}$ are closed).

If an operator P(x) mapping E into E_1 is additive and homogeneous, i.e., $P(x + y) = P(x) + P(y)$, $P(\lambda x) = \lambda P(x)$, it is called a linear operator. With addition of operators and multiplication of operators by a scalar defined in the natural way, the linear operators form a linear space. A linear operator which is continuous at one point is continuous. Moreover, a linear operator is continuous if and only if it is bounded in the neighborhood of any point (for this reason, continuous linear operators are often referred to as bounded linear operators).

A most important special case of a linear operator is a linear functional, i.e., a real-valued linear function on E. We usually denote linear functionals by f(x), g(x) or (f, x), (g, x), etc. The set of all continuous linear functionals on E is a linear space which is denoted by E' and known as the dual space.

If P is a linear operator mapping E into E_1, the adjoint operator is a linear operator P^* mapping E_1' into E', defined by $(P^*f, x) = (f, Px)$ for any f in E_1' and any x in E. It is clear that if P is continuous then so is P^*.

The following simple observation will be useful in the sequel.

Lemma 2.1. A linear functional which is nonnegative on an open set is continuous.

Indeed, if $f(x) \geq 0$ on an open set Q, then, by our previous remarks, there exists a symmetric neighborhood U of zero such that $x_0 + U \subset Q$ for some $x_0 \in Q$. Then $f(x_0 + x) \geq 0$, $f(x_0 - x) \geq 0$ for all $x \in U$, i.e., $|f(x)| \leq f(x_0)$ for $x \in U$, so that f(x) is bounded (and therefore continuous).

We recall the definition of linear manifold, subspace, hyperplane. A set L is called a linear manifold if, for any two vectors $x \in L$, $y \in L$, the vector $x + \alpha(y - x)$ is in L for any $\alpha \in R^1$. The closure of a linear manifold is also a linear manifold. A linear manifold containing the point 0 is called a subspace. A linear manifold defined by a nonzero linear functional, $L = \{x: f(x) = \alpha\}$, is called a hyperplane. If the functional f(x) in this definition is continuous, the hyperplane is closed. Every hyperplane H is either closed or everywhere dense (i.e., either $\overline{H} = H$ or $\overline{H} = E$). A set A situated on one side of a closed hyperplane (i.e., $A = \{x: f(x) \leq \alpha\}$, $f \in E'$) is called a closed half-space, and the set $B = \{x: f(x) < \alpha\}$, $f \in E'$, is called an open half-space. The following lemmas are well known.

Lemma 2.2. Any finite-dimensional linear manifold (i.e., a set

$$L = \{x_0 + \sum_{i=1}^{n} \lambda_i (x_i - x_0)\},$$

where $x_0, x_1, \ldots, x_n$ are fixed and $\lambda_1, \ldots, \lambda_n$ are arbitrary numbers) is closed.

Lemma 2.3. Let $f_1, f_2 \in E'$, $Q_1 = \{x: f_1(x) = 0\}$, $Q_2 = \{x: f_2(x) = 0\}$ and $Q_1 \subset Q_2$. Then either $f_2 = 0$ (i.e., $Q_2 = E$) or $f_1 = \lambda f_2$, $\lambda \neq 0$ (i.e., $Q_1 = Q_2$).

The principal topological linear spaces to be considered in this book will be Banach spaces (or B-spaces). These are linear space provided with a norm $\|x\|$ satisfying the usual conditions, and complete in this norm. A basis about zero in a B-space is the family of balls $U_\varepsilon = \{x: \|x\| < \varepsilon\}$, $\mathcal{A} = \{U_\varepsilon, \varepsilon > 0\}$.

Hilbert spaces constitute a special case of Banach spaces. A Hilbert space is provided with an inner product (x, y), defined for each pair of elements. In a Hilbert space, $E = E'$, i.e., the linear functionals are in one-to-one correspondence with the elements $y \in E$; this correspondence is defined by $f(x) = (x, y)$ for all $x \in E$.

We recall the most important Banach and Hilbert spaces; these will be used frequently in the sequel.

1. Finite-dimensional euclidean space R^n. The elements of this space are the n-vectors $x = (x_1, \ldots, x_n)$; the norm (denoted in this case by $|x|$) is defined by

$$|x| = \sqrt{\sum_{i=1}^{n} x_i^2}.$$

R^n is a Hilbert space, with $(x, y) = \sum_{i=1}^{n} x_i y_i$. Every linear operator P mapping R^m into R^n is uniquely determined by a matrix $P = (P_{ij})$, $i = 1, \ldots, n$, $j = 1, \ldots, m$, $y = Px$, where $Px \in R^n$ is understood as the product of the matrix P and the vector [$m \times 1$ matrix] $x \in R^m$; the adjoint operator, mapping R^n into R^m, is determined by the transpose P^*.

2. The space $L_p^{(n)}(a, b)$, $1 \leq p < \infty$ (denoted briefly by L_p). This is the space of n-tuples of measurable functions $x(t) = (x_1(t), \ldots, x_n(t))$ defined on [a, b]

and with summable p-th power. The norm is defined by

$$\| x \| = \left(\int_a^b | x(t)|^p \, dt \right)^{1/p} = \left(\int_a^b \left(\sum_{i=1}^n x_i^2(t) \right)^{p/2} dt \right)^{1/p}.$$

L_p is a Banach space. For $1 < p < \infty$, $E' = L_q$, where $\frac{1}{p} + \frac{1}{q} = 1$, i. e., if $f \in E'$, there exists a function $f(t) \in L_q^{(n)}$ such that

$$f(x) = \int_a^b (f(t), x(t)) \, dt = \int_a^b \left(\sum_{i=1}^n f_i(t) x_i(t) \right) dt.$$

When $p = 2$, L_p is a Hilbert space, with

$$(x, y) = \int_a^b (x(t), y(t)) \, dt.$$

3. The space $L_\infty^{(n)}(a, b)$ (or simply L_∞) is the space of all n-tuples of essentially bounded functions on [a, b], with norm

$$\| x \| = \operatorname*{ess\,sup}_{a \leqslant t \leqslant b} | x(t) |$$

where ess sup denotes the smallest number λ such that $|x(t)| \leq \lambda$ for almost all $a \leq t \leq b$. L_∞ is a Banach space. The dual space possesses a fairly complicated structure. At the same time, $L_\infty = L_1'$.

4. The space $C^{(n)}(a, b)$ (or simply C) is the space of n-tuples of functions continuous on [a, b], with norm

$$\| x \| = \max_{a < t \leqslant b} | x(t) | .$$

C is a Banach space, and its dual space is the space of measures on [a, b], i. e., for every $f \in C'$ there exists an n-dimensional measure $d\mu(t)$, with support [a, b], such that

$$f(x) = \int_a^b (x(t), d\mu(t)) = \int_a^b \sum_{i=1}^n x_i(t)\, d\mu_i(t).$$

We now return to the general theory of topological linear spaces.

Our next subject is a class of sets which will play an essential role in the sequel — the class of convex sets.

A set A in a linear space E is said to be convex if, together with any two points x, $y \in A$, it contains the entire segment [x, y]. In other words, if $x \in A$, $y \in A$ then $\lambda x + (1 - \lambda)y \in A$ for all $0 \leq \lambda \leq 1$. Elementary examples of convex sets are: linear manifold, segment, half-space (closed or open), ball in a Banach space.

The following properties of convex sets are obvious. The intersection of any number of convex sets is convex. If A, B are convex, then $\alpha A + \beta B$ is convex for any $\alpha, \beta \in R^1$. The image and preimage of a convex set under a linear mapping are convex sets. If E is a topological linear space and A is convex, then A^0 and $\bar{A}$ are convex sets. Given an arbitrary set A, we define its convex hull $\langle A \rangle$ as the smallest convex set containing A. It is readily seen that

$$\langle A \rangle = \{x : x = \sum_{i=1}^n \lambda_i x_i,\ x_i \in A,\ \lambda_i \geq 0,\ \sum_{i=1}^n \lambda_i = 1,\ 1 \leq n < \infty\}.$$

The convex hull of a closed set need not be closed. Below we shall need the following easily proved propositions.

Lemma 2.4. If $x \in A$, $y \in A^0$ and A is a convex set, then

$$\lambda x + (1 - \lambda) y \in A^0, \quad 0 \leq \lambda < 1.$$

Lemma 2.5. If A is a convex set and $A^0 \neq \emptyset$, then $\bar{A} = \overline{(A^0)}$, $A^0 = (\bar{A})^0$.

A topological linear space is said to be locally convex if it has a basis about zero consisting entirely of convex sets. All the standard function spaces (including all Banach spaces) are locally convex. Locally convex spaces are important for the reason that they possess the so-called separation properties, to be studied in Lecture 3.

Up to now we have regarded the dual E' simply as a linear space. We shall now define various topological structures on E', and also consider another topology in E (the weak topology).

Recall first that if τ_1 and τ_2 are two topologies in a linear space E (giving topological linear spaces E_{τ_1} and E_{τ_2}), then the topology τ_1 is said to be weaker than the topology τ_2 (or: τ_2 is stronger than τ_1) if any set which is open in τ_1 is also open in τ_2. Obviously, a necessary and sufficient condition for this state of affairs is that for every $U \in \mathcal{A}_{\tau_1}$ there exist $V \in \mathcal{A}_{\tau_2}$ such that $V \subset U$, i.e., every element of a basis about zero in τ_1 is contained in an element of a basis about zero in τ_2. Two topologies are said to be equivalent if their classes of open sets coincide. In particular, if a topology is defined in an n-dimensional space by any norm (for example,

$$|x| = \max_{1 \leqslant i \leqslant n} |x_i|, \quad |x| = (\sum_{i=1}^{n} |x_i|^p)^{1/p}, \quad 1 \leqslant p < \infty \quad \text{etc.}),$$

then all these topologies are equivalent (in particular, they are equivalent to the euclidean topology of R^n). Analogous remarks hold for the spaces $C^{(n)}$ and $L_p^{(n)}$ for any finite-dimensional norm $|x(t)|$.

If τ_1 is weaker than τ_2, then $E'_{\tau_1} \subset E'_{\tau_2}$ — every linear functional which is continuous in the topology τ_1 is also continuous in τ_2. However, the duals may be equal without the topologies being equivalent, though the classes of convex sets in both topologies will then coincide.

The strong topology of the space E' is given by the following basis about zero:

$$U = \{f \in E' : |f(x)| < \varepsilon \text{ for all } x \in A\},$$

where A ranges over all bounded sets in E and ε over all positive numbers. With this topology, the space E' will be denoted by E^*. If E is a Banach space, the topology thus defined is equivalent to the topology defined by the norm $\|f\| = \sup_{\|x\| \leq 1} |f(x)|$, and E^* is also a Banach space.

The E-<u>topology</u> of E' (or the <u>weak* topology</u>) is defined by the following basis about zero:

$$U = \{f \in E' : |f(x)| < \varepsilon \text{ for all } x \in A\},$$

where A now ranges over all finite sets (sets consisting of finitely many elements) in E and $0 < \varepsilon < \infty$. This is the weakest topology in which all linear functionals on E' of the type $x(f) = f(x)$ are continuous.

Besides the original topology of E, which we shall call the <u>strong</u> topology, we can define another topology — the <u>weak topology</u> or E'-<u>topology</u> of E, which has the following basis about zero:

$$U = \{x \in E : |f(x)| < \varepsilon \text{ for all } f \in A\},$$

where A ranges over all finite sets in E', $0 < \varepsilon < \infty$. This topology is the weakest one in which all functionals $f \in E'$ are continuous.

Finally, if E' is now taken as the original space, we can define a <u>weak topology</u> for it too (i. e., the $E^{*\prime}$-<u>topology</u> of E'), by the following basis:

$$U = \{f \in E' : |g(f)| < \varepsilon \text{ for all } g \in A\},$$

where A ranges over all finite sets in $E^{*\prime}$ ($E^{*\prime}$ is the set of all continuous linear functionals on E^*) and $0 < \varepsilon < \infty$.

We now consider some properties of these topologies. The space E' is locally convex in both weak and weak* topologies. In a finite-dimensional space, all topologies (strong, weak, weak*) are equivalent. In the general case, these topologies are distinct, and the relations among them correspond to their designations (the strong topology is stronger than the weak topology). The weak* topology is weaker than the weak topology of E' (and therefore any weakly closed set in E' is weakly* closed), but these topologies are equivalent in a reflexive Banach space. We recall that a Banach space is reflexive if $E^{*\prime} = E$ (i.e., to each continuous linear functional c on E^* corresponds an element $x \in E$ such that $(c, f) = (f, x)$ for all $f \in E$, and this correspondence is an isomorphism). In particular, every Hilbert space is reflexive. The L_p spaces $(1 < p < \infty)$ are reflexive, but the spaces L_1, L_∞ and C are not. The sets of continuous linear functionals in the strong and weak topologies of E or E' are the same. It follows that any convex set which is closed in the strong topology is also closed in the weak topology (this statement is of course generally false for non-convex sets).

To conclude this lecture, we recall some important definitions relating to the concept of compactness. A set A is said to be bicompact if, from every covering of A by open sets, one can extract a finite covering. A bicompact set is necessarily closed, and its image under any continuous mapping is also bicompact. Hence the important statement: a functional (not necessarily linear) which is continuous on a bicompact set assumes its maximum and minimum on the set. A set A in a topological space E is said to be conditionally compact if every countable sequence $x^n \in A$, $n = 1, 2, \ldots$, contains a subsequence converging to a point $x^0 \in E$; the set is compact if the point x^0 is in A. It turns out that a set in a metric space (in particular, a Banach space) is bicompact if and only if it is compact (or, equivalently, conditionally compact and closed). In a finite-dimensional space, every bounded set is conditionally compact. If E is a Banach space, then any weakly* closed bounded set in E^* is also weakly* bicompact. In a locally convex space, any closed convex set is weakly closed. These two statements and the fact that the weak and weak* topologies of a reflexive space are equivalent imply that a bounded closed convex set in a reflexive space is weakly bicompact. A few other important properties of bicompact sets are worthy of mention.

Lemma 2.6. Let A and B be closed sets in a topological linear space, and let B be bicompact. Then A + B is closed. If L_1, L_2 are closed subspaces and L_2 is finite-dimensional, then $L_1 + L_2$ is closed.

Lemma 2.7. The closed convex hull $\langle \bar{A} \rangle$ of a weakly bicompact set A in a Banach space is weakly bicompact.

The last definition that we need from the general theory of topological linear spaces is that of <u>direct product</u> (or simply <u>product</u>) of spaces. Let E_1, E_2 be two topological linear spaces. Consider the set of all pairs (x_1, x_2) with $x_1 \in E_1$, $x_2 \in E_2$. Introducing the natural linear operations

$$(x_1, x_2) + (x_1', x_2') = (x_1 + x_1', x_2 + x_2'), \quad \lambda (x_1, x_2) = (\lambda x_1, \lambda x_2),$$

and the basis about zero consisting of the pairs (U_1, U_2) with $U_1 \in \mathcal{A}_1$, $U_2 \in \mathcal{A}_2$, we get a new topological linear space with elements $x = (x_1, x_2)$, which is called the direct product of E_1 and E_2, denoted by $E_1 \times E_2$. The dual space is

$$E' = \{f = (f_1, f_2),\ f_1 \in E_1',\ f_2 \in E_2'\}, \text{ and } f(x) = f_1(x_1) + f_2(x_2).$$

In particular, if E_1 and E_2 are Banach spaces, then $E = E_1 \times E_2$ is also a Banach space: the norm of the element $x \in E$, $x = (x_1, x_2)$, $x_1 \in E_1$, $x_2 \in E_2$, is

$$\| x \| = (\| x_1 \|^2_{E_1} + \| x_2 \|^2_{E_2})^{1/2}.$$

The definition of the direct product of any finite number of spaces is analogous.

Lecture 3

HAHN-BANACH THEOREM

We shall now discuss the Hahn-Banach Theorem, which is the basic tool in the subsequent analysis. We shall not present proofs of the theorems, since they can be found in the extensive literature on functional analysis.

The Hahn-Banach Theorem can be given many formulations, but each of them is usually quite easy to derive from the others. We begin with the analytic (rather than geometric) versions of the theorem, which are concerned with the extension of a linear functional.

Theorem 3.1. Let $F(x)$ be a functional defined on a linear space E, which is convex and positively homogeneous (i.e., $F(\lambda x + (1-\lambda)y) \leq \lambda F(x) + (1-\lambda)F(y)$, $0 \leq \lambda \leq 1$, and $F(\lambda x) = \lambda F(x)$, $\lambda \geq 0$). Let $\bar{f}(x)$ be a linear functional defined on a subspace $L \subset E$ such that $\bar{f}(x) \leq F(x)$ for $x \in L$. Then there exists an extension of the functional $\bar{f}$ to the entire space E, possessing the same property, i.e., there exists a linear functional f defined on E such that $f(x) = \bar{f}(x)$ for $x \in L$ and $f(x) \leq F(x)$ for all $x \in E$.

It is important to note that the space in this formulation of the theorem is not assumed to be a topological space. In normed spaces the role of the functional $F(x)$ is usually played by $F(x) = \alpha \|x\|, \alpha > 0$.

This theorem has various corollaries on the existence of nonzero linear functionals. We present one of these.

Theorem 3.2. Let L be a closed subspace of a Banach space E, $x \notin L$. Then there exists $f \in E'$ such that $f(x) = 1$, $f(y) = 0$ for all $y \in L$. In particular, for each $x \neq 0$ there exists $f \in E'$ such that $\|f\| = 1$, $f(x) = \|x\|$.

In other words, there exist "sufficiently many" linear functionals in order to separate the points of a Banach space.

We now proceed to the geometric formulations of the Hahn-Banach Theorem, which are usually known as separation theorems. Let E be a topological linear space, A and B two sets in E. A nonzero continuous linear functional f is said to separate (strongly separate) A and B if there exists a number α such that $f(x) \geqslant \alpha$ for $x \in A$, $f(x) \leqslant \alpha$ for $x \in B$ ($f(x) > \alpha$, $x \in A$; $f(x) < \alpha$, $x \in B$). The closed hyperplane $H = \{x: f(x) = \alpha\}$ is then called a separating hyperplane for A and B, and the sets A and B themselves are said to be separable (Fig. 4). It is clear that if A and B are separable open sets, then they are also strongly separable. We now present the fundamental separation theorem.

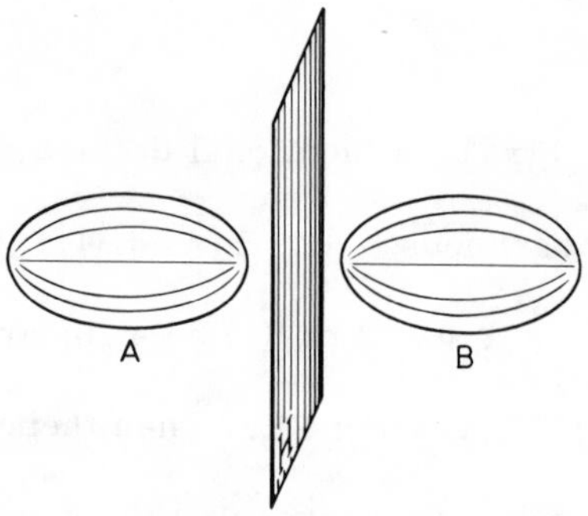

Fig. 4.

Theorem 3.3. Any two disjoint convex sets, one of which contains an interior point, are separable.

Trivial examples in a one-dimensional space show that the convexity assumption is essential. It is more difficult to prove the more profound statement that the assumption of the existence of an interior point cannot be relaxed.

The separation theorem is valid for closed convex sets only under the additional assumption that one of the sets is bicompact.

Theorem 3.4. If A and B are disjoint closed convex sets in a locally convex space and A is bicompact, then A and B are strongly separable.

Example 3.1.

$$E = R^2,\ A = \left\{x = (x_1, x_2):\ x_1 \geqslant 0,\ x_2 \geqslant \frac{1}{x_1}\right\},$$

$$B = \{x = (x_1, x_2):\ x_2 = 0\}.$$

Then A and B are convex and closed, but they are not strongly separable (though they are separable).

Example 3.2.

$$E = C(0,1),\ A = \{x:\ |x(t)| \leqslant 1,\ 0 \leqslant t \leqslant 1\},$$

$$B = \{x: \int_0^{1/2} x(t)\,dt - \int_{1/2}^{1} x(t)\,dt = 1\}.$$

Thus, A is the unit ball and B a closed hyperplane; hence A and B are closed and convex. It is easily seen that $A \cap B = \emptyset$, since any continuous function in B must satisfy the relation $\max_{0 \leqslant t \leqslant 1} |x(t)| > 1$. Now, no hyperplane H can strongly separate A and B: such a hyperplane would have to be parallel to B, and therefore satisfy the relation

$$f(x) = \int_0^{1/2} x(t)\,dt - \int_{1/2}^{1} x(t)\,dt = \alpha,$$

but if $|\alpha| > 1$ or $\alpha = -1$ the sets A and B lie on one side of $H = \{x:\ f(x) = \alpha\}$, and if $|\alpha| < 1$ the hyperplane H cuts A. Note that the continuous linear functional f(x) of this example does not assume a maximum on the unit sphere; this is related to the fact that the space C is not reflexive. It can be shown that all continuous linear functionals assume a maximum on the sphere if and only if the space is reflexive.

Theorem 3.4 has several immediate corollaries valid for a locally convex space.

Corollary 1. A point can be strongly separated from any convex closed (and by Theorem 3.3, also open) set which does not contain it.

Corollary 2. If $x, y \in E$, then there exists $f \in E'$ such that $f(x) \neq f(y)$ (i.e., in a locally convex space there are sufficiently many continuous linear functionals in order to separate points).

Corollary 3. Every closed convex set A is the intersection of the half-spaces that contain it.

Indeed, let $B = \bigcap H$, $H = \{x: f(x) \leq \alpha\}$, where the intersection extends over all $f \in E'$, $-\infty < \alpha < \infty$, such that $f(x) \leq \alpha$ for all $x \in A$. It is obvious that $A \subset B$. Let $B \neq A$. Take $x_0 \in B$, $x_0 \notin A$. By Corollary 1, there exist $f_0 \in E'$, $\alpha_0 \in R^1$, such that $f_0(x) < \alpha_0$ for $x \in A$, $f_0(x_0) > \alpha_0$. Then $x_0 \notin H_0 = \{x: f_0(x) \leq \alpha_0\}$, which contradicts the definition of B.

Applying Theorem 3.4 to sets in spaces with the weak topologies and using the conditions for weak bicompactness set forth in Lecture 2, we arrive at the following results.

Theorem 3.5. Let E be a reflexive Banach space, A and B disjoint closed convex sets in E, with A bounded. Then A and B are strongly separable.

Theorem 3.6. Let E be a topological linear space, A a weakly* closed convex set in E', $f_0 \notin A$. Then there exists a point $x \in E$ such that $f_0(x) < f(x)$ for all $f \in A$.

Such sets in E', which are separable from any point not contained in them by functionals of the type $x(f) = f(x)$, are sometimes called regularly convex sets; thus Theorem 3.6 gives conditions for regular convexity.

Lecture 4

SUPPORTING HYPERPLANES AND EXTREMAL POINTS

A nonzero continuous linear functional f is said to be a supporting (sometimes: tangent) functional for a set $A \subset E$ at $x_0 \in A$ if $f(x) \geqslant f(x_0)$ for all $x \in A$. Under these conditions, the closed hyperplane $H = \{x: f(x) = f(x_0)\}$ is called a supporting hyperplane for A at the point x_0. The geometric sense of a supporting hyperplane is quite simple: the set A lies on one side of the hyperplane and cuts it in one point x_0 (Fig. 5).

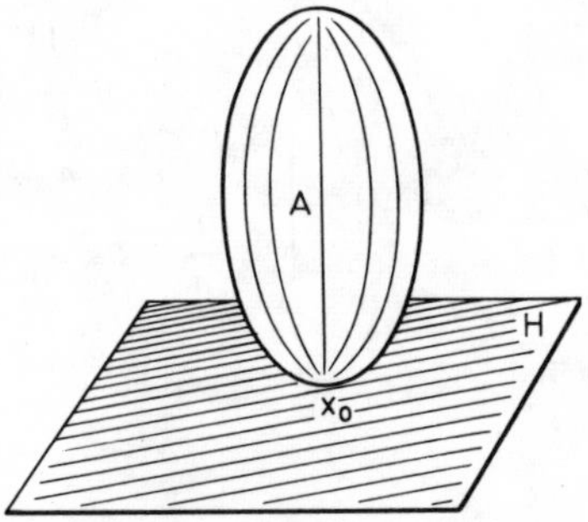

Fig. 5.

The properties of supporting functionals will be considered somewhat later; for the moment we shall examine the question of their existence. First, it is quite obvious that no supporting functional can exist at an interior point of the set. If the set A is a convex body (i.e., a closed convex set with interior points), then, as will follow from the theorem stated below, it has a supporting functional at each boundary point.

Theorem 4.1. A convex body has a supporting functional at each boundary point.

Indeed, if $x_0 \in A$ is a boundary point of A, then $x_0 \notin A^0$. The set A^0 is non-empty, convex and open; therefore, by Theorem 3.3, it can be separated from the point x_0. Hence there exist a functional $f \in E'$ and a number α such that $f(x) > \alpha$ for all $x \in A^0$, and $f(x_0) \leqslant \alpha$. Now, by Lemma 2.5, in the case of a convex body we have $\overline{A^0} = A$, and so $f(x) \geqslant \alpha$ for all $x \in A$, in particular, $f(x_0) \geqslant \alpha$. Thus $f(x_0) = \alpha$, and the linear functional f is a supporting functional for A at x_0.

On the other hand, if the set A is convex but has no interior points, it may not have a supporting functional at its boundary points. We consider an important counterexample.

Example 4.1. Let $A = \{x \in L_2(0, 1): |x(t)| \leqslant 1 \text{ for almost all } 0 \leqslant t \leqslant 1\}$.

This set is clearly convex, and is readily seen to be closed. It has no interior points. In fact, if $x_0 \in A$, consider the set of points $x^n = x_0 + y^n$, where

$$y^n(t) = \begin{cases} 3, & 0 \leqslant t \leqslant \dfrac{1}{n}, \\ 0, & \dfrac{1}{n} < t \leqslant 1. \end{cases}$$

Then $x^n \notin A$ for all $n \geqslant 1$, but

$$\| x^n - x_0 \| = \| y^n \| = \frac{3}{\sqrt{n}} \to 0 \text{ as } n \to \infty,$$

and this means that x_0 is a boundary point of A.

Thus every point $x_0 \in A$ is a boundary point of A. However, it is easily seen that the set A does not have a supporting functional at each $x_0 \in A$. Take $x_0(t) \equiv 0$. Then, if $c(t) \in L_2$ is a supporting functional for A at the point x_0, we must have $(c, x_0) \leqslant (c, x)$ for all $x \in A$, in particular, for $x(t) = -\text{sign}\, c(t)$:

$$0 \leqslant -\int_0^1 c(t) \,\text{sign}\, c(t)\, dt = -\int_0^1 |c(t)|\, dt,$$

and this is possible only if $c(t) \equiv 0$. In exactly the same way, one can show that A has no supporting functional at all points $x_0 \in A$ such that $|x_0(t)| < 1$ on a set of positive measure. On the other hand, there is a supporting functional at all points $x_0 \in A$ such that $|x_0(t)| \equiv 1$ (one need only take $c(t) = -\text{sign } x_0(t)$). There are other examples of convex sets (even compacta) which have no supporting functional at their their boundary points.

We now introduce another important concept. A point x_0 of a set A in a linear spaee is called an <u>extremal point</u> of A (or a <u>vertex</u> of A) if it is not an interior point of any segment whose endpoints are in A. In other words, x_0 is a vertex if there exist no points $x_1 \in A$, $x_2 \in A$, $x_1 \neq x_2$, such that $x_0 = \lambda x_1 + (1 - \lambda)x_2$ for some $0 < \lambda < 1$. Thus, the point x_0 in Fig. 6 is a vertex of A, but the points x_1, x_2 are not.

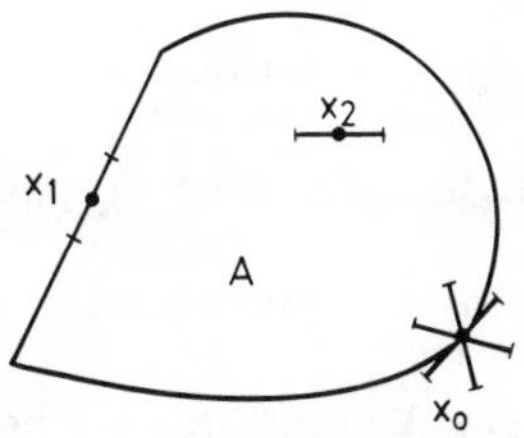

Fig. 6.

It is obvious that an open set has no vertices. On a two-dimensional plane, the extremal points of a closed polygon coincide with its vertices in the sense of the usual high-school definition; for a closed circle, the entire boundary consists of extremal points. In general, a bounded closed set in a finite-dimensional space always has extremal points (see Theorem 4.2 below), but this may not be true for an unbounded set (for example, a closed half-space has no extremal points). In an infinite-dimensional space, even a bounded closed set need not have vertices. For example, in the space C(0, 1), the set A of all functions x(t) such that $|x(t)| \leq 1$, $0 \leq t \leq 1$, $x(0) = 0$ (i.e., the intersection of a ball and a hyperplane), has no vertices. In fact, if $x_0 \in A$, then there exists $\varepsilon > 0$ such that $|x_0(t)| \leq 1/2$ for $0 \leq t \leq \varepsilon$, and so, if we take

$$h(t) = \begin{cases} \frac{t}{\varepsilon}, & 0 \leqslant t \leqslant \frac{\varepsilon}{2}, \\ 1 - \frac{t}{\varepsilon}, & \frac{\varepsilon}{2} < t \leqslant \varepsilon, \\ 0, & \varepsilon < t \leqslant 1 \end{cases}$$

and set $x_1(t) = x_0(t) + h(t)$, $x_2(t) = x_0(t) - h(t)$, then $x_1 \in A$, $x_2 \in A$, $x_0 = (x_1 + x_2)/2$. However, it can be shown that this situation cannot arise in bicompacta.

Theorem 4.2. A bicompact set in a locally convex topological linear space has extremal points.

Moreover, in a certain sense a bicompact convex set is generated by its vertices. In rigorous terms, we have the following

Theorem 4.3 (Krein-Mil'man). A bicompact convex set in a locally convex topological linear space is the closed convex hull of the set of its vertices.

The extremal points of a set A stand in a direct relation to extremum problems. Recall that $x_0 \in A$ is said to be a minimum point of a functional f(x) on A if $f(x_0) \leqslant f(x)$ for all $x \in A$ (in particular, if f(x) is a linear functional it is a supporting functional for A at x_0), and a strict minimum point if the inequality is strict for all $x \neq x_0$.

Theorem 4.4. Let f(x) be a linear functional, A a set in the linear space E. Then, if x_0 is a strict minimum point of f(x) on A, x_0 is an extremal point of A. Moreover, if A is a convex bicompact set, f(x) is continuous and E a locally convex topological linear space, then at least one minimum point of f(x) on A is an extremal point.

Proof. Let x_0 be a strict minimum point. Suppose that it is not an extremal point, i.e., there exist $x_1 \in A$, $x_2 \in A$, $x_1 \neq x_2$, $0 < \lambda < 1$, such that $x_0 = \lambda x_1 + (1 - \lambda)x_2$. Then

$$f(x_1) > f(x_0),\ f(x_2) > f(x_0);$$

therefore

$$f(x_0) = \lambda f(x_1) + (1-\lambda) f(x_2) > \lambda f(x_0) + (1-\lambda) f(x_0) = f(x_0),$$

which is impossible.

We now prove the second part of the theorem. Let B be the subset of A containing all minimum points of f(x) on A (B is nonempty since f(x) is continuous and A bicompact — see Lecture 2). Since f(x) is linear and A convex, the set B is convex; since f(x) is continuous and A closed, B is closed. Since B A, it follows that B is a bicompact convex set. By Theorem 4.2, it has an extremal point x_0. We claim that x_0 is also an extremal point of A. Let $x_1 \in A$, $x_2 \in A$, $x_1 \neq x_2$, $x_0 = \lambda x_1 + (1-\lambda)x_2$, $0 < \lambda < 1$. Since x_0 is a minimum point, it follows that

$$f(x_0) = f^* = \min_{x \in A} f(x),$$

while

$$f(x_1) \geqslant f^*,\ f(x_2) \geqslant f^*.$$

Thus the equalities $f^* = f(x_0) = \lambda f(x_1) + (1-\lambda) f(x_2)$ imply that $f(x_1) = f(x_2) = f^*$, but then $x_1 \in B$, $x_2 \in B$, which is absurd, since x_0 is an extremal point of B. This completes the proof.

Theorem 4.4 is the basis for the methods of linear programming. In fact, this theorem implies that, if we are searching for the minimum of a linear function on a bounded convex set in a finite-dimensional space, we may restrict the search to the vertices of the set.

Lecture 5

CONES, DUAL CONES

We now proceed to a systematic study of cones — sets of a special type which will play an important role in the sequel.

A set K in a linear space E is called a *cone with apex at* 0 if $\lambda K = K$ for any $\lambda > 0$ (i.e., if $x \in K$ implies that $\lambda x \in K$ for all λ 0).

If K is a cone with apex at 0, then the set $x_0 + K$ is called a *cone with apex at* x_0. We shall call a cone *sharp* or *blunt* according to whether it does or does not contain its apex. If the cone does not contain any straight line (i.e., it follows from $x \in K$, $x \neq 0$ that $-x \notin K$), it is said to be *proper* (or *salient*). It is clear that any intersection of cones with a common apex, or any set of the form $\alpha K_1 + \beta K_2$ (where K_1 and K_2 are cones) is also a cone. In a topological linear space, the interior of a cone, its closure and its convex hull are all cones. A cone which is a convex set is called a *convex cone*. A convex cone with apex at 0 may be defined as a set which, together with any two points x, y, contains all points of the form $\alpha x + \beta y$, $\alpha > 0$, $\beta > 0$.

The general separation theorems (Lecture 3) can be sharpened when the sets involved are cones.

L e m m a 5.1. Let K be a cone with apex at x_0, f(x) a linear functional such that $f(x) \geqslant \alpha$ for $x \in K$. The $f(x) \geqslant f(x_0)$ for $x \in K$.

Indeed, let $y \in K$ and $f(y) < f(x_0)$. By the definition of a cone, the entire ray $x_0 + t(y - x_0)$, $t > 0$, must lie in K, but $f(x_0 + t(y - x_0)) = f(x_0) + t(f(y) - f(x_0))$ for sufficiently large t, which contradicts the assumption of the lemma.

The geometrical meaning of this lemma is that, if the cone lies on one side of a hyperplane, then it must lie on one side of the parallel hyperplane through its apex (see Fig. 7).

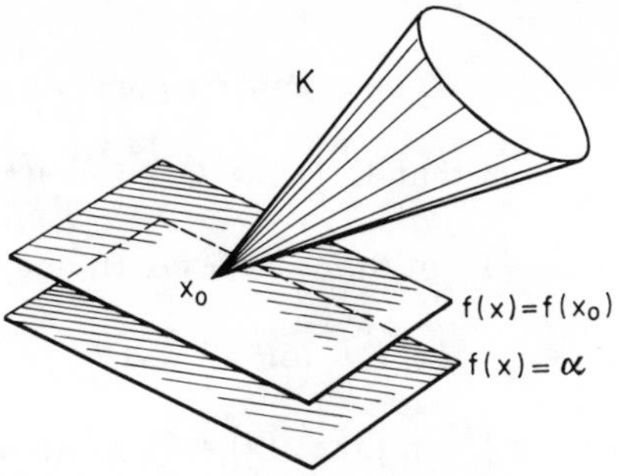

Fig. 7.

Lemma 5.1 implies the following corollaries.

a) If a cone K and a set A are separable by a hyperplane, then they are also separable by the parallel hyperplane through the apex of the cone (i.e., if $f(x) \geqslant \alpha$ for $x \in K$, $f(x) \leqslant \alpha$ for $x \in A$, then $f(x) \geqslant f(x_0)$ for $x \in K$, $f(x) \leqslant f(x_0)$ for $x \in A$, since $f(x_0) \geqslant \alpha$).

b) A closed convex cone in a locally convex space can be separated from any ray issuing from its apex and not passing through the cone.

In fact, let x_0 be the apex of the cone, x_1 a point on the ray, $x_1 \neq x_0$. Then $x_1 \notin K$ (otherwise the entire ray would lie in K), and by Theorem 3.4 x_1 and K are separable: there exist $f \in E'$ and α such that $f(x) \geqslant \alpha$ for $x \in K$, $f(x_1) \leqslant \alpha$. By Corollary a), we have $f(x) \geqslant f(x_0)$ for $x \in K$, $f(x_1) \leqslant f(x_0)$. Therefore, for any point $y = x_0 + t(x_1 - x_0)$ on the ray we have $f(y) = f(x_0) + t(f(x_1) - f(x_0)) \leqslant f(x_0)$, i.e., the hyperplane $f(x) = f(x_0)$ separates the cone and the ray.

It is also readily seen that the hyperplane can be so chosen that it strongly separates arbitrary points of the cone and the ray (other than the apex of the cone).

c) If two cones K_1 and K_2 with apex at 0 are separable by a hyperplane, then the latter must pass through 0. In particular, if a cone with apex at 0 and a subspace L are separable by a hyperplane, then L is contained in the hyperplane.

We now prove an important theorem on extensions of a positive linear functional.

Theorem 5.1 (Krein). Let K be a convex cone with apex at 0, containing interior points, L a subspace such that $K^0 \cap L \neq \emptyset$. Let $\overline{f}(x)$ be a linear functional on L such that $\overline{f}(x) \geq 0$ on $K \cap L$. Then there is a continuous linear functional f(x) on E such that $f(x) = \overline{f}(x)$ for $x \in L$, $f(x) \geq 0$ for all $x \in K$.

Proof. If $\overline{f}(x) \equiv 0$ on L, we can take $f(x) \equiv 0$; we therefore assume that $\overline{f}$ is not identically zero. Let $Q_1 = \{x \in L: \overline{f}(x) = 0\}$. Then Q_1 is convex and nonempty ($0 \in Q_1$). Moreover, $Q_1 \cap K^0 = \emptyset$, for if $x_0 \in Q_1 \cap K^0$, then $\overline{f}(x_0) = 0$, but since $x_0 \in L \cap K^0$, there is a point $x_1 \in L \cap K$ in the neighborhood of x_0 such that $\overline{f}(x_1) < 0$, contradicting the fact that $\overline{f}(x)$ is nonnegative on $L \cap K$. Hence there exists a hyperplane separating Q_1 and K^0. In other words, there exists a linear functional g(x) (which may be assumed continuous — see Lemma 2.1) such that $g(x) > 0$ on K^0, $g(x) = 0$ on Q_1 (this follows from Corollary c) of the preceding lemma). Consider the two linear functionals $\overline{f}(x)$ and g(x) on the subspace L, and their sets of zeros $Q_1 = \{x \in L: \overline{f}(x) = 0\}$, $Q_2 = \{x \in L: g(x) = 0\}$. $Q_1 \subset Q_2$, since $g(x) = 0$ on Q_1. Thus there are only two possibilities: either $Q_1 = Q_2$, or $Q_2 = L$ (Lemma 2.3). But the second case is impossible, since for any point $x_0 \in L \cap K^0$ we have $g(x_0) > 0$. Consequently, by Lemma 2.3 $\overline{f}(x) = \lambda g(x)$, $\lambda \neq 0$, on L. λ must be positive, since $\overline{f}(x)$ and g(x) have the same sign on $K \cap L$. Thus $f(x) = \lambda g(x)$ is the required extension of the functional $\overline{f}(x)$.

We now introduce the important concept of the dual cone. Let E be a topological linear space, K a cone in E with apex at 0. The <u>dual cone</u> $K^* \subset E'$ is defined as the set of all continuous linear functionals nonnegative on K (i.e., $K^* = \{f \in E': f(x) \geq 0 \text{ for all } x \in K\}$). It follows directly from the definition that K^* is a sharp convex cone with apex at 0.

Example 5.1.

$$E = R^2,\ K = \{x = (x_1, x_2): x_1 \geqslant 0,\ x_2 \geqslant 0\},\ K^* = K.$$

Example 5.2.

$$E = R^2,\ K = \{x: (a_1, x) \geqslant 0,\ (a_2, x) \geqslant 0\},\ a_1 \in R^2,\ a_2 \in R^2,$$

$$K^* = \{f: f = \lambda_1 a_1 + \lambda_2 a_2,\ \lambda_1 \geqslant 0,\ \lambda_2 \geqslant 0\} \text{ (Fig. 8).}$$

More interesting examples of dual cones will be presented later. We now note some properties of dual cones. Throughout the sequel, K_1, K_2, etc. will be cones with apex at 0 in a locally convex space.

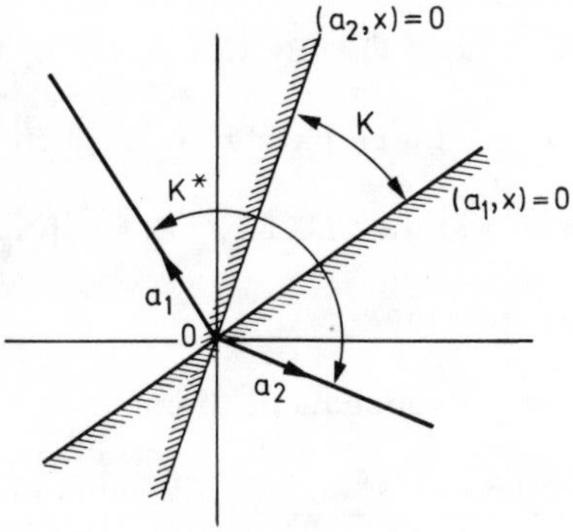

Fig. 8.

Lemma 5.2. K^* is weakly* closed.

Indeed, let $f_0 \notin K^*$. Then there exists $x_0 \in K$ such that $f_0(x_0) < 0$. Therefore, for all $f \in U$, where $U = \{f: |f(x_0)| < -f_0(x_0)\}$, we have $f(x_0) < 0$, i.e., $U \cap K^* = \emptyset$. But U is a weak* neighborhood of f_0, and so the complement of K^* is weakly* open.

Lemma 5.3. $K^* = (\overline{K})^* = \langle\overline{K}\rangle^*$ (throughout this section, the bar will denote weak closure), i.e., the dual cone is unchanged if the original cone is replaced by its weak closure or even by the weak closure of its convex hull.

In fact, if $f(x) \geqslant 0$ on K, then, since f(x) is linear it is nonnegative on $\langle K \rangle$, and since it is continuous it remains nonnegative on the weak closure $<\overline{K}>$.

Lemma 5.4. $(\bigcup_{\alpha\in A} K_\alpha)^* = \bigcap_{\alpha\in A} K_\alpha^*$ for any index set A.

This is an obvious consequence of the definition.

Lemma 5.5. If $K_1 \subset K_2$, then $K_1^* \supset K_2^*$.

This is also obvious from the definition.

Set

$$K^{**} = \{x \in E: f(x) \geq 0 \text{ for all } f \in K^*\}.$$

Lemma 5.6. $K^{**} = <\overline{K}>$, i.e., K^{**} is the weak closure of the convex hull of K.

Indeed, let $Q = <\overline{K}>$. Then, by Lemma 5.3, $Q^* = K^*$. But if $x \in Q$, then $f(x) \geq 0$ for all $f \in Q^*$, and hence it follows from the definition of K^{**} that $x \in K^{**}$. Thus $Q \subset K^{**}$. Now let $x_0 \in K^{**}$, $x_0 \notin Q$. By Corollary b) to Lemma 5.1, the point x_0 can be separated from Q, i.e., there exists $f \in E'$ such that $f(x) \geq 0$ for all $x \in Q$ and $f(x_0) < 0$. Therefore, $f \in Q^*$, and so $f \in K^*$. But if $f(x_0) < 0$, $f \in K^*$, then $x_0 \notin K^{**}$, contradicting the assumption.

A very important question in subsequent discussions will be that of the <u>cone dual to the intersection of a given set of cones.</u> It turns out that, in some cases, $(\bigcap_{\alpha\in A} K_\alpha)^* = <\bigcup_{\alpha\in A} K_\alpha^*>$. A rigorous formulation of this statement follows. Note that the equality $<\bigcup_{\alpha\in A} K_\alpha^*> = \sum_{\alpha\in A} K_\alpha^*$ always holds, where $\sum_{\alpha\in A} K_\alpha^*$ denotes the the set of all finite sums $f_{\alpha_1} + \ldots + f_{\alpha_n}$, $f_{\alpha_i} \in K_{\alpha_i}^*$, $\alpha_i \in A$, A being an arbitrary index set.

Lemma 5.7. $(\bigcap_{\alpha\in A} K_\alpha)^* \supset \sum_{\alpha\in A} K_\alpha^*$. This is obvious from the definition.

Lemma 5.8. Let K_α, $\alpha \in A$, be weakly closed and convex cones. Then $(\bigcap_{\alpha\in A} K_\alpha)^* = \overline{<\sum_{\alpha\in A} K_\alpha^*>}$, i.e., the dual cone of the intersection is the weakly* closed convex hull of the sum of dual cones.

Proof.

Let $Q = \overline{\langle \sum_{\alpha\in A} K_\alpha^* \rangle}$. Then $Q^* = \langle \bigcup_{\alpha\in A} K_\alpha^* \rangle^* = (\bigcup_{\alpha\in A} K_\alpha^*)^*$, by Lemma 5.3. It follows from Lemmas 5.4 and 5.6 that

$$(\bigcup_{\alpha\in A} K_\alpha^*)^* = \bigcap_{\alpha\in A} K_\alpha^{**} = \bigcap_{\alpha\in A} \overline{\langle K_\alpha \rangle} = \bigcap_{\alpha\in A} K_\alpha .$$

Thus $Q^* = \bigcap_{\alpha\in A} K_\alpha$, and so $Q^{**} = (\bigcap_{\alpha\in A} K_\alpha)^*$. But by Lemma 5.6 we have $Q^{**} = Q$, and hence $Q = (\bigcap_{\alpha\in A} K_\alpha)^*$.

Corollary. If the cones K are weakly closed and convex, while $\sum_{\alpha\in A} K_\alpha^*$ is weakly* closed, then

$$(\bigcap_{\alpha\in A} K_\alpha)^* = \sum_{\alpha\in A} K_\alpha^* .$$

Lemma 5.9. Let K be a convex cone with apex at 0, L a subspace such that $K^0 \cap L \neq \emptyset$. Then $(K\cap L)^* = K^* + L^*$.

This property is essentially another version of Theorem 5.1. In fact, if $\bar{f} \in (K\cap L)^*$, then $\bar{f}(x) \geqslant 0$ on $K\cap L$, and by Theorem 5.1 there exists $f \in K^*$ such that $\bar{f}(x) = f(x)$ on L. Then $\bar{f} = f + (\bar{f} - f)$, $f \in K^*$, $\bar{f} - f \in L^*$ (since $(f - \bar{f})(x) = 0$ for $x \in L$), so that $\bar{f} \in K^* + L^*$. The inverse inclusion follows from Lemma 5.7.

Lemma 5.10. Let $K_1, \ldots, K_n$ be convex open cones, $\bigcap_{i=1}^n K_i \neq \emptyset$. Then

$$(\bigcap_{i=1}^n K_i)^* = \sum_{i=1}^n K_i^* .$$

Proof. We shall employ the following geometric construction. Consider the topological product $\tilde{E} = \underbrace{E \times E \times \ldots \times E}_{n}$, where the points $\tilde{x} \in \tilde{E}$ have the form

$$\tilde{x} = (x_1, x_2, \ldots, x_n),\ x_i \in E,$$

and $\tilde{E}' = E' \times E' \times \ldots \times E'$ is the set of functionals on $\tilde{E}$:

$$\tilde{f} = (f_1, \ldots, f_n),\ \tilde{f} \in \tilde{E}',\ f_i \in E',\ \tilde{f}(\tilde{x}) = \sum_{i=1}^{n} f_i(x_i)$$

(see Lecture 2).

Consider the set $K = \{\tilde{x}: x_i \in K_i,\ i = 1, \ldots, n\}$ in $\tilde{E}$. This set is an open convex cone (since it is the direct product of open convex cones). Consider another set in $\tilde{E}$, $L = \{\tilde{x}: x_1 = x_2 = \ldots = x_n = x \in E\}$. L is a linear subspace of $\tilde{E}$. It follows from the condition $\bigcap K_i \neq \emptyset$ that $L \cap K \neq \emptyset$. Consider the linear functional $\overline{f}(\tilde{x})$ defined on the subspace L by $\overline{f}(\tilde{x}) = f(x)$, where $\tilde{x} = (x, x, \ldots, x) \in L$, $x \in E$, and $f \in (\bigcap K_i)^*$. Then $\overline{f}$ is nonnegative on $L \cap K$ (since $\overline{f}(\tilde{x}) = f(x)$, but if $\tilde{x} = (x, \ldots, x) \in L \cap K$, then $x \in \bigcap K_i$, and $f \in (\bigcap K_i)^*$). Theorem 5.1 therefore applies, and hence there exists $\tilde{f} \in \tilde{E}'$ which is nonnegative on K and coincides with $\overline{f}$ on L. Let $\tilde{f} = (f_1, \ldots, f_n)$. Then on the subspace L we have $\tilde{f}(\tilde{x}) = \sum_{i=1}^{n} f_i(x) = \overline{f}(\tilde{x}) = f(x)$, that is, $f = \sum_{i=1}^{n} f_i$, where $f \in (\bigcap K_i)^*$. And since $\tilde{f}(\tilde{x}) \geqslant 0$ for $\tilde{x} \in K$, it follows that $\sum_{i=1}^{n} f_i(x_i) \geqslant 0$ for $x_i \in K_i$, $i = 1, \ldots, n$, and hence $f_i(x_i) \geqslant 0$ for $x_i \in K_i$, i.e., $f_i \in K_i^*$, $i = 1, \ldots, n$. Thus, we see that if $f \in (\bigcap K_i)^*$, then there exist $f_i \in K_i^*$ such that $f = \sum_{i=1}^{n} f_i$, in other words, $(\bigcap K_i)^* \subset \sum K_i^*$. The inverse inclusion follows from Lemma 5.7.

R e m a r k. We outline another proof of Lemma 5.10. As follows from Lecture 2, $\overline{\bigcap K_i} = \bigcap \overline{K}_i$, since the K_i are convex open sets and $\bigcap K_i$ is nonempty. Thus, by Lemma 5.3, $(\bigcap K_i)^* = (\overline{\bigcap K_i})^* = (\bigcap \overline{K}_i)^*$. We may therefore assume that the K_i are weakly closed. One can now use the Corollary to Lemma 5.8, first proving directly that $\sum K_i^*$ is weakly* closed (in so doing one must use Lemma 2.6). This method of proof, though more instructive, is somwhat less intuitive.

The fundamental result of this lecture can now be formulated without difficulty.

Lemma 5.11 (Dubovitskii-Milyutin). Let $K_1, \ldots, K_n, K_{n+1}$ be convex cones with apex at 0, where $K_1, \ldots, K_n$ are open. Then $\bigcap_{i=1}^{n+1} K_i = \emptyset$ if and only if there exist linear functionals $f_i \in K_i^*$, $i = 1, \ldots, n+1$, not all zero, such that

$$f_1 + f_2 + \ldots + f_n + f_{n+1} = 0.$$

Proof. Necessity. Let $\bigcap_{i=1}^{n} K_i = K \neq \emptyset$, $K_{n+1} \cap K = \emptyset$. Since K is open, we can apply the separation theorem 3.3, which, together with Lemma 5.1, implies that there exists $f \in E'$, $f \neq 0$, such that $f(x) \geqslant 0$ for $x \in K$, $f(x) \leqslant 0$ for $x \in K_{n+1}$. By Lemma 5.10, it follows from $f \in K^*$ that $f = \sum_{i=1}^{n} f_i$, $f_i \in K_i^*$, $i = 1, \ldots, n$. Set $f_{n+1} = -f_n$; then $f_{n+1} \in K_{n+1}^*$, $f_{n+1} \neq 0$ and $f_1 + \ldots + f_n + f_{n+1} = 0$. But if $\bigcap_{i=1}^{n} K_i = \emptyset$, then there exists $1 \leqslant s < n$ such that $K = \bigcap_{i=1}^{s} K_i \neq \emptyset$, $\bigcap_{i=1}^{s+1} K_i = \emptyset$. Applying the result just proved (with n replaced by s), we get functionals $f_i \in K_i^*$, $i = 1, \ldots, s+1$, $f_{s+1} \neq 0$, such that $f_1 + \ldots + f_{s+1} = 0$. Taking $f_{s+2} = \ldots = f_{n+1} = 0$, we get the required functionals $f_1, \ldots, f_{n+1}$.

Sufficiency. Let $f_1 + \ldots + f_{n+1} = 0$, $f_i \neq 0$, while $\bigcap_{i=1}^{n+1} K_i \neq \emptyset$. There exists $x_0 \in K_i$, $i = 1, \ldots, n+1$. At the same time, for some $1 \leqslant j \leqslant n$ there must be a functional $f_j \neq 0$ (since otherwise we would have $f_{n+1} = -\sum_{i=1}^{n} f_i = 0$), and so $f_j(x_0) > 0$ (since K_j is open). But then $0 = (f_1 + \ldots + f_n + f_{n+1})(x_0) \geqslant f_j(x_0) > 0$, and this is absurd.

Lecture 6

NECESSARY EXTREMUM CONDITIONS

(EULER-LAGRANGE EQUATION)

We have now prepared sufficient auxiliary material to enable us to proceed to analysis of extremum problems.

Let F(x) be a functional (generally nonlinear), defined in a neighborhood of a point x_0 in a locally convex topological linear space E. We shall assume that the variable x must satisfy constraints of two types: $x \in Q_i$, i = 2, . . . , n, where Q_i are sets with nonempty interior, and $x \in Q_{n+1}$, where Q_{n+1} has no interior points. In the usual formulation, the sets Q_i, i = 1, . . . , n, are given by inequality constraints, and Q_{n+1} by a system of equality constraints, so that, as a rule, Q_{n+1} is a manifold of lower dimension than the space. The problem is to determine a local minimum of the functional F(x) on $\bigcap_{i=1}^{n+1} Q_i$, in other words, to find a point $x_0 \in Q$ such that

$$F(x_0) = \min_{Q \cap U} F(x),$$

where U is some neighborhood of x_0. What conditions must hold at the point x_0?

In order to answer this question, we must specify the class of functionals and sets to be investigated.

In classical analysis and the calculus of variations, the functionals are usually assumed to be differentiable. However, this class of functionals is too narrow, and we shall extend it considerably.

We shall call a vector h a direction of decrease of the functional F(x) at the point x_0 if there exist a neighborhood U of the vector h and a number $\alpha = \alpha(F, x_0, h)$, $\alpha < 0$, such that for all $0 < \varepsilon < \varepsilon_0$ and any $\bar{h} \in U$,

$$F(x_0 + \varepsilon \bar{h}) \leqslant F(x_0) + \varepsilon\alpha. \qquad (6.1)$$

Lemma 6.1. The directions of decrease generate an open cone K with apex at 0.

Proof. If h is a direction of decrease, then λh, $\lambda > 0$, is also a direction of decrease (simply replace U by λU, ε_0 by ε_0/λ, and α by $\alpha\lambda$). Therefore, K is a cone with apex at 0. Now, if $h \in K$ and $\bar{h} \in U$ (in the notation of the definition), then $\bar{h} \in K$ (since inequality (6.1) will hold with the same U, α, ε_0), and hence K is open.

F(x) is said to be a regularly decreasing functional if the set of its directions of decrease at a point x_0 is convex.

Similar definitions will now be introduced for the constraints of our problem.

Given inequality constraints, we shall say that a vector h is a feasible direction for Q at a point x_0 if there exists a neighborhood U of the vector h such that, for all $0 < \varepsilon < \varepsilon_0$ and all $\bar{h} \in U$, the vectors $x_0 + \varepsilon \bar{h}$ are in Q (Fig. 9).

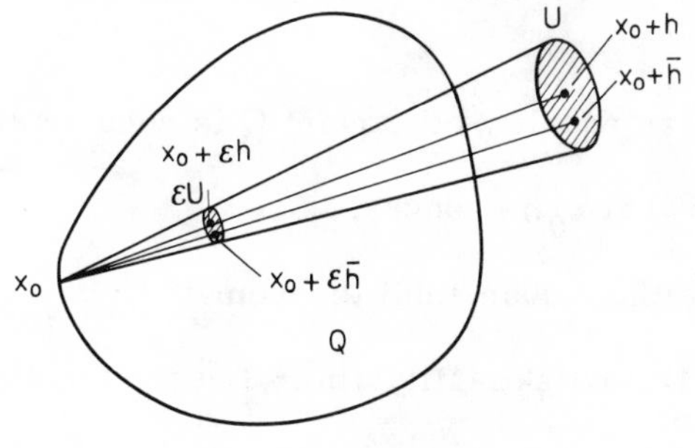

Fig. 9.

Lemma 6.2. The feasible directions generate an open cone K with apex at 0.

Proof. If $h \in K$, then $\lambda h \in K$ for $\lambda > 0$ (replace U by λU, ε_0 by ε_0/λ). Moreover, if $h \in K$, then all $\bar{h} \in U$ are also in K.

We shall say that an inequality constraint Q is regular at a point x_0 if the cone of feasible directions for Q at x_0 is convex.

For equality constraints (i.e., manifolds with no interior points), the set of feasible directions (in the sense of the original definition) is empty. We therefore introduce a slightly different concept for this type of constraint.

A vector h is said to be a tangent direction to Q at a point x_0 if, for any $0 < \varepsilon < \varepsilon_0$, there exists a point $x(\varepsilon) \in Q$ such that, if we set $x(\varepsilon) = x_0 + \varepsilon h + r(\varepsilon)$ (Fig. 10), then the vector $r(\varepsilon) \in E$ has the following property: for any neighborhood U of zero, $\frac{1}{\varepsilon} r(\varepsilon) \in U$ for all small $\varepsilon > 0$ (in a Banach space this is replaced by the simpler condition $\|r(\varepsilon)\| = o(\varepsilon)$). It is easily seen that the tangent directions also generate a cone K with apex at 0. However, this cone is in general neither closed nor open. In the majority of cases, K is a half-space. Note that every feasible direction is also a tangent direction, but the converse is false.

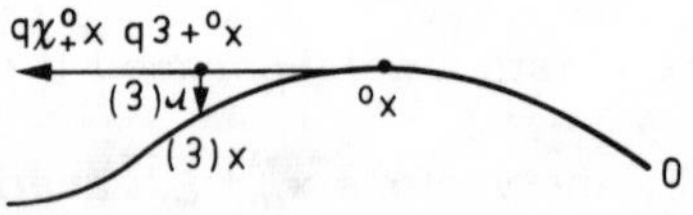

Fig. 10.

We shall say that an equality constraint Q is regular at a point x_0 if the cone of tangent directions to Q at x_0 is convex.

We now formulate the fundamental theorem.

Theorem 6.1 (Dubovitskii-Milyutin). Let the functional F(x) assume a local minimum on $Q = \bigcap_{i=1}^{n+1} Q_i$ at a point $x_0 \in Q$. Assume that F(x) is regularly decreasing at x_0, with directions of decrease K_0; the inequality constraints Q_i, $i = 1, \ldots, n$, are regular at x_0, with feasible directions K_i; the equality constraint Q_{n+1} is also regular at x_0, with tangent directions K_{n+1}. Then there exist continuous linear functionals f_i, $i = 0, 1, \ldots, n+1$, not all identically zero, such that $f_i \in K_i^*$, $i = 0, 1, \ldots, n+1$, which satisfy the Euler-Lagrange equation:

$$f_0 + f_1 + \ldots + f_n + f_{n+1} = 0. \tag{6.2}$$

Proof. We shall first prove that a necessary condition for the functional to have a minimum at x_0 is $\bigcap_{i=0}^{n+1} K_i = \emptyset$ (i.e., no direction of decrease of the functional F(x) can be feasible for all constraints). Suppose that this is false, so that there exists $h \in K_i$, $i = 0, \ldots, n+1$. By the definition of K_i, $i = 0, \ldots, n$, there exists a neighborhood U of the vector h such that, whenever $0 < \varepsilon < \varepsilon_0$, any vector $x_0 + \varepsilon \bar{h}$, $\bar{h} \in U$, lies in $\bigcap_{i=1}^{n} Q_i$ and satisfies inequality (6.1). Now consider the vector $x(\varepsilon) = x_0 + \varepsilon h + r(\varepsilon) \in Q_{n+1}$ as in the definition of tangent directions, and let ε_1 be such that $\frac{1}{\varepsilon} r(\varepsilon) \in U - h$, or $\bar{h}(\varepsilon) = h + \frac{1}{\varepsilon} r(\varepsilon) \in U$, for $0 < \varepsilon < \varepsilon_1$. Then, whenever $0 < \varepsilon < \min(\varepsilon_0, \varepsilon_1)$, the vectors $x(\varepsilon) = x_0 + \varepsilon \bar{h}(\varepsilon)$ are, on the one hand, in $\bigcap_{i=1}^{n} Q_i$, and, on the other, in Q_{n+1}. In other words, these vectors satisfy all the constraints. But they also satisfy inequality (6.1):

$$F(x_0 + \varepsilon \bar{h}(\varepsilon)) \leqslant F(x_0) + \varepsilon\alpha < F(x_0),$$

which contradicts the assumption that x_0 is a minimum point. Thus $\bigcap_{i=0}^{n+1} K_i = \emptyset$. Now, by definition, $K_0, K_1, \ldots, K_n$ are convex open cones, and K_{n+1} is a convex cone. Lemma 5.11 is therefore applicable, and this implies the required result.

Remark 1. Later we shall show that our extremum condition is a generalization of the rule of Lagrange multipliers in classical analysis and the Euler equation in the calculus of variations; this justifies our use of the name "Euler-Lagrange equation".

Remark 2. If there are no equality constraints ($Q_{n+1} = E$), the extremum condition can be strengthened, extending the set of directions of decrease and the set of feasible directions. Condition (6.1) may be replaced by the weaker inequality

$$F(x_0+\varepsilon h) \leqslant F(x_0)+\varepsilon\alpha, \quad \alpha<0,$$

and in the definition of feasible direction it suffices to require that $x_0 + \varepsilon h \in Q_i$, $0 < \varepsilon < \varepsilon_0$.

Remark 3. It is sometimes important to ensure that $f_0 \neq 0$. An examination of the proof of Theorem 6.1 shows that a sufficient condition for this to hold is that there exist at least one feasible direction, i.e., that $\bigcap_{i=1}^{n+1} K_i \neq \emptyset$. The analogous conclusion is valid for any f_i, $i = 0, \ldots, n+1$.

It follows from the Dubovitskii-Milyutin theorem that, if we wish to determine necessary conditions for an extremum in some specific problem, we must learn how to solve the following problems:

1. <u>To determine the directions of decrease.</u>
2. <u>To determine the feasible directions.</u>
3. <u>To determine the tangent directions.</u>
4. <u>To construct the dual cones.</u>

We shall now proceed to attack these problems; each will be considered in one of the following lectures. In this analysis, the distinguishing feature of the extremum condition as given by equation (6.2) is that each functional or constraint can be analyzed in isolation. Thus the analysis of each functional or set need be carried out only once and for all, and the results can then be utilized for any extremal problem in which it figures.

Lecture 7

DIRECTIONS OF DECREASE

In this lecture we shall see how to calculate the directions of decrease for various functionals.

A functional F(x) in a linear space E is said to have a derivative $F'(x_0, h)$ at a point x_0 in the direction h if

$$\lim_{\varepsilon \to +0} \frac{F(x_0 + \varepsilon h) - F(x_0)}{\varepsilon} = F'(x_0, h). \qquad (7.1)$$

If $E = R^1$ (so that F(x) is a function of one variable), then the existence of $F'(x_0, h)$ is equivalent to the existence of the right (for $h > 0$) or left (for $h < 0$) derivative of F(x) at the point x_0. The directional derivative has the following obvious properties (assuming throughout that $F_i(x)$, $i = 0, 1, 2$, are differentiable at x_0 in the direction h):

1) If $F_2(x) = \alpha F_0(x) + \beta F_1(x)$, then

   $F_2'(x_0, h) = \alpha F_0'(x_0, h) + \beta F_1'(x_0, h)$.

2) $F_0'(x_0, \lambda h) = \lambda F_0'(x_0, h)$ for $\lambda > 0$.

As we shall see below, directional derivatives furnish a convenient tool for constructing the cone of directions of decrease.

Throughout this lecture, K will denote the cone of directions of decrease for the functional F(x) at the point x_0.

Theorem 7.1. If $h \in K$ and $F'(x_0, h)$ exists, then $F'(x_0, h) < 0$.

Proof. It follows from (6.1) that $F(x_0 + \varepsilon h) \leqslant F(x_0) + \varepsilon \alpha$, and therefore

$$\lim_{\varepsilon \to +0} \frac{F(x_0 + \varepsilon h) - F(x_0)}{\varepsilon} \leqslant \alpha < 0.$$

The fact that the converse is valid for important classes of functionals is much less trivial.

Theorem 7.2. Let E be a Banach space and let F(x) satisfy a Lipschitz condition in a neighborhood of x_0 (i.e., there exist $\varepsilon_0 > 0$ and $\beta > 0$ such that $|F(x_1) - F(x_2)| \leqslant \beta \|x_1 - x_2\|$ for all $\|x_1 - x_0\| \leqslant \varepsilon_0$, $\|x_2 - x_0\| \leqslant \varepsilon_0$) and $F'(x_0, h) < 0$. Then $h \in K$.

Proof. Let $F'(x_0, h) = -\delta < 0$. It then follows from (7.1) that $F(x_0 + \varepsilon h) \leqslant F(x_0) - \varepsilon\delta/2$ for $0 < \varepsilon < \varepsilon_1$. Let $\bar{h}$ be an arbitrary vector such that $\|h - \bar{h}\| \leqslant \delta/4$. Then, whenever

$$0 < \varepsilon < \min(\varepsilon_0, \varepsilon_1)$$

we have

$$F(x_0 + \varepsilon\bar{h}) \leqslant F(x_0 + \varepsilon h) + \beta \|\varepsilon\bar{h} - \varepsilon h\| \leqslant$$
$$\leqslant F(x_0) - \frac{\varepsilon\delta}{2} + \frac{\varepsilon\delta}{4} = F(x_0) - \frac{\varepsilon\delta}{4},$$

that is, condition (6.1) holds with $\alpha = -\delta/4$, and therefore $h \in K$.

Our next task is to determine when a functional is regularly decreasing (i.e., when the cone K is convex). To this end we introduce the important concept of a convex functional.

A functional F(x) defined on a linear space E is said to be convex if, for any $x, y \in E$ and any $0 \leqslant \lambda \leqslant 1$,

$$F(\lambda x + (1 - \lambda) y) \leqslant \lambda F(x) + (1 - \lambda) F(y). \tag{7.2}$$

The following properties of convex functionals are direct consequences of the definition.

Lemma 7.1. If $F_i(x)$, $i = 1, \dots, n$, are convex functionals and $\alpha_i \geqslant 0$,

$i = 1, \ldots, n$, then $\sum_{i=1}^{n} \alpha_i F_i(x)$ is also a convex functional. If $F_s(x)$, $s \in S$, are convex functionals (where S is an arbitrary index set), then $\sup_{s \in S} F_s(x)$ is also a convex functional. If F(x) is a convex functional, then the sets $\{x: F(x) < \lambda\}$, $\{x: F(x) \leqslant \lambda\}$ are convex for any λ.

We can now answer the question concerning regularly decreasing functionals.

Theorem 7.3. Let E be a Banach space. Let F(x) satisfy a Lipschitz condition in a neighborhood of the point $x_0 \in E$. Assume that F(x) is differentiable at x_0 in any direction, and that $F'(x_0, h)$ is convex as a functional of h. Then F(x) is regularly decreasing at x_0, and $K = \{h: F'(x_0, h) < 0\}$.

Proof. The fact that $K = \{h: F'(x_0, h) < 0\}$ follows from Theorems 7.1 and 7.2. Since $F'(x_0, h)$ is convex in h, it follows from Lemma 7.1 that $K = \{h: F'(x_0, h) < 0\}$ is convex.

We now describe two important special cases in which $F'(x_0, h)$ is convex in h. In so doing, we shall be able to weaken some of the assumptions of Theorem 7.3.

Theorem 7.4. Let F(x) be a convex continuous functional in a topological linear space E. Then F(x) is differentiable at any point in any direction, $F(x_0 + h) \geqslant F(x_0) + F'(x_0, h)$, F(x) is regularly decreasing at any point, and $K = \{h: F'(x_0, h) < 0\}$.

Proof. Let $x_0 \in E$, $h \in E$ be fixed, and consider the function of one variable $\varphi(\varepsilon) = F(x_0 + \varepsilon h)$. Since F(x) is convex and continuous, this function is also convex and continuous. Hence, for any $0 < \lambda < 1$ we have $\varphi(\lambda\varepsilon) \leqslant \lambda\varphi(\varepsilon) + (1-\lambda)\varphi(0)$, i.e., for $\varepsilon > 0$,

$$\frac{\varphi(\lambda\varepsilon) - \varphi(0)}{\lambda\varepsilon} \leqslant \frac{\varphi(\varepsilon) - \varphi(0)}{\varepsilon}.$$

Hence it follows that $\dfrac{\varphi(\varepsilon) - \varphi(0)}{\varepsilon}$ is monotone decreasing for $\varepsilon \longrightarrow +0$.

Furthermore, for $\varepsilon > 0$ we have

$$\varphi(0) \leqslant \frac{\varepsilon}{\varepsilon+1}\varphi(-1) + \frac{1}{\varepsilon+1}\varphi(\varepsilon), \quad \frac{\varphi(\varepsilon)-\varphi(0)}{\varepsilon} \geqslant \varphi(-1)-\varphi(0),$$

so that $\dfrac{\varphi(\varepsilon)-\varphi(0)}{\varepsilon}$ is bounded below for $\varepsilon > 0$. Thus, $\lim\limits_{\varepsilon\to+0}\dfrac{\varphi(\varepsilon)-\varphi(0)}{\varepsilon}$ exists. But it follows from the definition of the directional derivative that

$$F'(x_0, h) = \lim_{\varepsilon\to+0}\frac{\varphi(\varepsilon)-\varphi(0)}{\varepsilon}.$$

Thus, the derivative exists at each point in an arbitrary direction. Moreover, the fact that $\dfrac{\varphi(\varepsilon)-\varphi(0)}{\varepsilon}$ is monotone decreasing implies that

$$\varphi(1)-\varphi(0) \geqslant \lim_{\varepsilon\to+0}\frac{\varphi(\varepsilon)-\varphi(0)}{\varepsilon},$$

and this in turn means that

$$F(x_0+h) \geqslant F(x_0) + F'(x_0, h).$$

Assuming now that $F'(x_0, h) < 0$, let us prove that $h \in K$. It follows from (7.1) that there exists a number $\varepsilon_0 > 0$ such that $F(x_0 + \varepsilon_0 h) \quad F(x_0)$; set

$$\delta = F(x_0) - F(x_0+\varepsilon_0 h) > 0.$$

Let U be a neighborhood of the point h such that, for all $\bar{h} \in U$,

$$|F(x_0+\varepsilon_0\bar{h}) - F(x_0+\varepsilon_0 h)| \leqslant \frac{\delta}{2}$$

(this is possible because $F(x)$ is continuous). Then $F(x_0 + \varepsilon_0\bar{h}) \leqslant F(x_0) - \delta/2$.

It now follows from the convexity condition for $0 < \varepsilon < \varepsilon_0$ that

$$F(x_0+\varepsilon\bar{h}) \leqslant \left(1-\frac{\varepsilon}{\varepsilon_0}\right)F(x_0)+\frac{\varepsilon}{\varepsilon_0}F_0(x_0+\varepsilon_0\bar{h}) \leqslant F(x_0)-\frac{\varepsilon\delta}{2\varepsilon_0},$$

that is, condition (6.1) holds with $\alpha = -\delta/2\varepsilon_0$. This means that $h \in K$. Together with Theorem 7.1, this implies that $K = \{h: F'(x_0, h) < 0\}$.

Finally, let us prove that K is convex. Let $h_1 \in K$, $h_2 \in K$; we must show that $h = \lambda h_1 + (1-\lambda)h_2 \in K$ for $0 < \lambda < 1$. We have

$$\frac{F(x_0+\varepsilon h)-F(x_0)}{\varepsilon} \leqslant$$
$$\leqslant \frac{\lambda[F(x_0+\varepsilon h_1)-F(x_0)]+(1-\lambda)[F(x_0+\varepsilon h_2)-F(x_0)]}{\varepsilon},$$

and therefore

$$F'(x_0, h) \leqslant \lambda F'(x_0, h_1) + (1-\lambda)F'(x_0, h_2).$$

But since $h_1 \in K$, $h_2 \in K$ implies that $F'(x_0, h_1) < 0$, $F'(x_0, h_2) < 0$, it follows that $F'(x_0, h) < 0$, i.e., $h \in K$. This completes the proof of the theorem.

We now consider another important class of functionals which are regularly decreasing at any point, whose cones of directions of decrease are easily determined. A functional F(x) defined on a Banach space E is said to be differentiable (or Fréchet-differentiable) at a point x_0 if there exists a linear functional $f \in E'$ such that, for all $h \in E$,

$$F(x_0+h) = F(x_0) + (f, h) + o(\|h\|). \qquad (7.3)$$

The linear functional f is denoted by $F'(x_0)$ or $DF(x_0)$, and called the derivative, Fréchet-derivative or gradient of the functional F(x) at the point x_0.

Note that if $E = R^n$ this definition is equivalent to that of a differentiable function of severable variables, and $F'(x)$ is the vector in R^n with coordinates

$$\left(\frac{\partial F(x)}{\partial x_1}, \dots, \frac{\partial F(x)}{\partial x_n}\right).$$

Lemma 7.2. If $F_1(x)$ and $F_2(x)$ are differentiable at x_0, then the functional $F(x) = \alpha_1 F_1(x) + \alpha_2 F_2(x)$ is also differentiable at x_0 and

$$F'(x_0) = \alpha_1 F_1'(x_0) + \alpha_2 F_2'(x_0).$$

If $F(x)$ is differentiable at x_0, then it is differentiable in any direction, and $F'(x_0, h) = (F'(x_0), h)$. If $F(x)$ is differentiable at x_0 and convex, then

$$F(x_0+h) \geqslant F(x_0) + (F'(x_0), h). \tag{7.4}$$

The last assertion of the lemma follows from the inequality

$$F(x_0+h) \geqslant F(x_0) + F'(x_0, h)$$

proved in Theorem 7.4. The remainder of the lemma is obvious.

We can now strengthen Theorem 7.3 for differentiable functionals.

Theorem 7.5. If $F(x)$ is differentiable at x_0, then $F(x)$ is regularly decreasing at x_0 and $K = \{h: (F'(x_0), h) < 0\}$.

Proof. It follows from Theorems 7.1, 7.2 and Lemma 7.2 that $K = \{h: (F'(x_0), h) < 0\}$. But $\{h: (F'(x_0), h) < 0\}$ is convex (for it is either a half-space or the empty set) and therefore $F(x)$ is regularly decreasing.

Thus, Theorems 7.3, 7.4 and 7.5 describe the class of regularly decreasing functionals and enable one to determine the directions of decrease using directional derivatives or derivatives. We now present some examples. In each example it will suffice to calculate the directional derivative and to apply the above-mentioned theorems.

Example 7.1. Consider the linear functional $F(x) = (f, x)$, where $f \in E'$. Then $F(x)$ is differentiable, $F'(x) \equiv f$ and $K = \{h: (f, h) < 0\}$ for any point x_0.

Example 7.2. Integral functional. $E = C^{(n)}(0, T)$, $x = x(t)$,

$$F(x) = \int_0^T \Phi(x(t), t)dt.$$

Let $\Phi(x, t)$ be continuous in x, t and differentiable with respect to x, with partial derivatives $\Phi_x(x, t)$ continuous in x, t. Then $F(x)$ is differentiable,

$$(F'(x_0), h) = \int_0^T (\Phi_x(x_0(t), t), h(t))dt$$

$$K = \{h \in C : \int_0^T (\Phi_x(x_0(t), t), h(t))\, dt < 0\}.$$

Indeed,

$$F(x_0+h)-F(x_0) = \int_0^T [\Phi(x_0(t)+h(t), t)-\Phi(x_0(t), t)]dt =$$

$$= \int_0^T (\Phi_x(x_0(t), t), h(t))\, td + \int_0^T \Big([\Phi_x(x_0(t)+\theta(t)h(t), t)-$$

$$- \Phi_x(x_0(t), t)], h(t) \Big)\, dt, \quad 0 \leqslant \theta(t) \leqslant 1;$$

But $\Phi_x(x, t)$ is continuous on $R^n \times R^1$, and therefore uniformly continuous on $M \times [0, T]$, where $M \subset R^n$, $M = \{x: |x| \leq \|x_0\| + \|h\|\}$, i.e., for every $\varepsilon > 0$ there exists $\delta > 0$ such that $|\Phi_x(x_1, t) - \Phi_x(x_2, t)| \leq \varepsilon$ for $x_1, x_2 \in M$, $|x_1 - x_2| \leq \delta$, $t \in [0, T]$. Since $x_0(t) \in M$, $x_0(t) + \theta(t)h(t) \in M$ for all $t \in [0, T]$, it now follows that

$$| \Phi_x(x_0(t)+\theta(t)h(t), t)-\Phi_x(x_0(t), t) | \leqslant \varepsilon$$

for $|h(t)| \leqslant \delta$, i.e., for $\|h\| \leqslant \delta$. Consequently,

$$\max_{0<t<T} | \Phi_x(x_0(t)+\theta(t)h(t), t)-\Phi_x(x_0(t), t) | \to 0$$

as $\|h\| \longrightarrow 0$, and this means that

$$| \int_0^T \Big([\Phi_x(x_0(t)+\theta(t)h(t), t)-\Phi_x(x_0(t), t)], h(t)\Big) dt | \leqslant$$
$$\leqslant T \max_{0<t<T} | \Phi_x(x_0(t)+\theta(t)h(t), t)-\Phi_x(x_0(t), t) | \cdot \| h \| = o(\| h \|).$$

Example 7.3. Integral function in the space $C \times L_\infty$. We slightly complicate the preceding example. Set

$$F(x, u) = \int_0^T \Phi(x(t), u(t), t)dt,$$

where $x \in C^{(n)}(0, T)$ and $u \in L_\infty^{(r)}(0, T)$, $\Phi(x, u, t)$ is continuous in x, u, t, differentiable with respect to x, u; $\Phi_x(x, u, t)$ and $\Phi_u(x, u, t)$ are continuous in x, u, t. Then

$$F'(x_0, u_0)(\bar{x}, \bar{u}) = \int_0^T [(\Phi_x(x_0(t), u_0(t), t), \bar{x}(t)) +$$
$$+(\Phi_u(x_0(t), u_0(t), t), \bar{u}(t))]dt,$$
$$K = \{\bar{x} \in C, \bar{u} \in L_\infty : F'(x_0, u_0)(\bar{x}, \bar{u}) < 0\}.$$

The proof is exactly the same as in the preceding example, since there we used only the boundedness of the functions x(t) and h(t) and never used their continuity.

We now consider some examples of functionals which are not smooth.

Example 7.4. The maximum of a finite set of functionals. Let

$$F(x) = \max_{1<i<n} F_i(x),$$

where $F_i(x)$, $i = 1, \ldots, n$, are functionals in a linear space E, differentiable at x_0

in the direction h. Then F(x) is differentiable at x_0 in the direction h, and

$$F'(x_0, h) = \max_{i \in I} F_i'(x_0, h), \quad I = \{i\colon F_i(x_0) = F(x_0)\}.$$

Proof. Let

$$M = \max_{1 \leqslant i \leqslant n} | F_i'(x_0, h) |, \quad \delta = \max_{i \notin I} (F(x_0) - F_i(x_0)) > 0.$$

Choose $\varepsilon_i > 0$ such that

$$\left| \frac{F_i(x_0 + \varepsilon h) - F_i(x_0)}{\varepsilon} - F_i'(x_0, h) \right| \leqslant M$$

for $0 < \varepsilon < \varepsilon_i$, and denote

$$\varepsilon_0 = \min \left\{ \varepsilon_1, \ldots, \varepsilon_n, \frac{\delta}{4M} \right\} > 0.$$

Then, if $0 < \varepsilon < \varepsilon_0$ and $i \in I$,

$$F_i(x_0 + \varepsilon h) \geqslant F_i(x_0) + \varepsilon F_i'(x_0, h) - \varepsilon M \geqslant F(x_0) - 2\varepsilon M > F(x_0) - \frac{\delta}{2},$$

and if $i \notin I$,

$$\begin{aligned} F_i(x_0 + \varepsilon h) &\leqslant F_i(x_0) + \varepsilon F_i(x_0, h) + \varepsilon M \leqslant \\ &\leqslant F_i(x_0) + 2\varepsilon M \leqslant F(x_0) - \frac{\delta}{2}, \end{aligned}$$

Therefore, for $0 < \varepsilon < \varepsilon_0$, we have

$$\begin{aligned} F(x_0 + \varepsilon h) &= \max_{1 \leqslant i \leqslant n} F_i(x_0 + \varepsilon h) = \max_{i \in I} F_i'(x_0 + \varepsilon h) = \\ &= \max_{i \in I} (F_i(x_0) + \varepsilon F_i'(x_0, h) + o_i(\varepsilon)) = F(x_0) + \varepsilon \max_{i \in I} F_i'(x_0, h) + o(\varepsilon). \end{aligned}$$

Hence

$$F'(x_0, h) = \lim_{\varepsilon \to +0} \left(\max_{i \in I} F_i'(x_0, h) + \frac{o(\varepsilon)}{\varepsilon} \right) = \max_{i \in I} F_i'(x_0, h).$$

We now derive an important corollary. If each functional $F_i(x)$ is either (i) convex and continuous or (ii) differentiable, then $F(x)$ is regularly decreasing at x_0 and $K = \{h: F_i'(x_0, h) < 0, \ i \in I\}$. Indeed, in either of these cases $F_i'(x_0, h)$ exists and is convex in h (for any i), and therefore (see Lemma 7.1) max $F_i'(x_0, h)$ is also convex in h, i.e., K is convex.

Example 7.5. Maximum of a function over a closed interval:

$$F(x) = \max_{0 \leq t \leq T} G(x(t), t),$$

where $x \in C^{(n)}(0, T)$. Let $G(x, t)$ be continuous in x, t, differentiable with respect to x, and $G_x(x, t)$ continuous in x, t. Then $F(x)$ is differentiable at any point in any direction, and

$$F'(x_0, h) = \max_{t \in R} (G_x(x_0(t), t), h(t)),$$
$$R = \{t \in [0, T]: G(x_0(t), t) = F(x_0)\}.$$

Moreover, $F(x)$ is regularly decreasing at any point and

$$K = \{h \in C: (G_x(x_0(t), t), h(t)) < 0, \quad t \in R\}.$$

Proof. First, reasoning exactly as in Example 7.2, we get

$$F(x_0+\varepsilon h) = \max_{0\leqslant t\leqslant T} (G(x_0(t)+\varepsilon h(t), t)) =$$
$$= \max_{0\leqslant t\leqslant T} [G(x_0(t), t)+\varepsilon(G_x(x_0(t), t), h(t))]+o(\varepsilon).$$

Since $G(x_0(t), t) \equiv F(x_0)$ for $t \in R$, it follows that

$$\max_{0\leqslant t\leqslant T} [G(x_0(t), t)+\varepsilon(G_x(x_0(t), t), h(t))] \geqslant \max_{t\in R} [G(x_0(t), t) +$$
$$+ \varepsilon(G_x(x_0(t), t), h(t))] = F(x_0)+\varepsilon \max_{t\in R} (G_x(x_0(t), t), h(t)).$$

Denote

$$S_\varepsilon = \{\tau\in [0, T]: G(x_0(\tau), \tau)+\varepsilon(G_x(x_0(\tau), \tau), h(\tau)) =$$
$$= \max_{0\leqslant t\leqslant T} (G(x_0(t), t)+\varepsilon(G_x(x_0(t), t), h(t)))\},$$
$$R_\delta = \{\tau\in [0, T]: \min_{t\in R} | t-\tau | \leqslant \delta\},$$

i. e., S_ε is the set of maximum points of the function $G(x_0(t), t) + \varepsilon (G_x(x_0(t), t), h(t))$, and R_δ is the δ-neighborhood of the set R. We claim that for any $\delta > 0$ there exists $\varepsilon_0 > 0$ so small that $S_\varepsilon \subset R_\delta$ for all $0 < \varepsilon < \varepsilon_0$. If this is not so, we can find a sequence of numbers $\varepsilon_i \longrightarrow 0$ and a sequence of points $t_{\varepsilon_i} \in S_{\varepsilon_i}$ such that

$$\min_{t\in R} | t_{\varepsilon_i} - t | \geqslant \delta > 0$$

for all i. The sequence t_{ε_i} contains a convergent subsequence $t_{\varepsilon_k} \longrightarrow t^*$. By the definition of S_{ε_k},

$$G(x_0(t_{\varepsilon_k}),t_{\varepsilon_k})+\varepsilon_k(G_x(x_0(t_{\varepsilon_k}),t_{\varepsilon_k}), h(t_{\varepsilon_k}))\geqslant G(x_0(t),t)+\varepsilon_k(G_x(x_0(t),t),h(t))$$

for all $t \in R$, and hence

$$G(x_0(t_{\varepsilon_k}), t_{\varepsilon_k})\geqslant F(x_0)-2\varepsilon_k \max_{0\leqslant t<T} | (G_x(x_0(t), t), h(t)) | .$$

But it follows from the definition of R that

$$G(x_0(t_{\varepsilon_k}), t_{\varepsilon_k}) \leqslant F(x_0),$$

and now, combining these inequalities, we get

$$\lim_{k\to\infty} G(x_0(t_{\varepsilon_k}), t_{\varepsilon_k}) = F(x_0).$$

Therefore, by the continuity of G(x, t) and $x_0(t)$,

$$G(x_0(t^*), t^*) = F(x_0),$$

i. e., $t^* \in R_0$. But this contradicts the assumption

$$\min_{t\in R} | t_{\varepsilon_k} - t | \geqslant \delta > 0.$$

Thus, we indeed have $S_\varepsilon \subset R_\delta$ for sufficiently small $\varepsilon > 0$. Therefore,

$$\max_{t\in S_\varepsilon} (G_x(x_0(t), t), h(t)) \leqslant \max_{t\in R_\delta} (G_x(x_0(t), t), h(t)).$$

Since $G_x(x, t)$, $x_0(t)$ and h(t) are continuous,

$$\lim_{\delta\to 0} \max_{t\in R_\delta} (G_x(x_0(t), t), h(t)) = \max_{t\in R} (G_x(x_0(t), t), h(t))$$

and hence

$$\varepsilon \max_{t\in S_\varepsilon} (G_x(x_0(t), t), h(t)) \leqslant \varepsilon \max_{t\in R} (G_x(x_0(t), t), h(t)) + o(\varepsilon).$$

Consequently,

$$\max_{0\leqslant t\leqslant T} [G(x_0(t), t) + \varepsilon (G_x(x_0(t), t), h(t))] =$$

$$= \max_{t\in S_\varepsilon} [G(x_0(t), t) + \varepsilon (G_x(x_0(t), t), h(t))] \leqslant$$

$$\leqslant \max_{t\in S_\varepsilon} G(x_0(t), t) + \varepsilon \max_{t\in S_\varepsilon} (G_x(x_0(t), t), h(t)) \leqslant$$
$$\leqslant F(x_0) + \varepsilon \max_{t\in R}(G_x(x_0(t), t), h(t)) + o(\varepsilon),$$

and this, together with the previously proved inequality, gives

$$\max_{0\leqslant t\leqslant T} [G(x_0(t), t) + \varepsilon(G_x(x_0(t), t), h(t))] = F(x_0) +$$
$$+ \varepsilon \max_{t\in R} (G_x(x_0(t), t), h(t)) + o(\varepsilon).$$

Substituting this into the expression for $F(x_0 + \varepsilon h)$, we finally get

$$F(x_0 + \varepsilon h) = F(x_0) + \varepsilon \max_{t\in R} (G_x(x_0(t), t), h(t)) + o(\varepsilon),$$

and this proves our formula for $F'(x_0, h)$.

Since

$$| F(x_1) - F(x_2) | = | \max G(x_1(t), t) - \max G(x_2(t), t) | \leqslant$$
$$\leqslant \max | G(x_1(t), t) - G(x_2(t), t) | =$$
$$= \max | (G_x(x_\theta(t), t), x_1(t) - x_2(t)) | \leqslant L \| x_1(t) - x_2(t) \| ,$$

where

$$L = \max_{\substack{|x|\leqslant\rho \\ 0\leqslant t\leqslant T}} G_x(x, t), \quad \rho = \max \{ \| x_1 \| , \| x_2 \| \},$$
$$x_\theta(t) = x_1(t) + \theta(t)(x_2(t) - x_1(t)), \quad 0 \leqslant \theta(t) \leqslant 1,$$

it follows that $F(x)$ satisfies a Lipschitz condition in any ball. Moreover, $F'(x_0, h)$ is convex in h (as the maximum of the functions $G_x(x_0(t), t)$, $h(t)$, which are convex in h; see Lemma 7.1). We may therefore apply Theorem 7.3, so that $K = \{h: F'(x_0, h) < 0\}$, proving the required formula for K.

Example 7.6. Maximum of a non-smooth function. It is not difficult to see that a slight modification of the arguments for the preceding example yields a similar result when $G(x, t)$ has only a directional derivative. We present the final result for one example only: $G(x, t) = |x|$, i. e.,

$$F(x) = \max_{0 \leqslant t \leqslant T} |x(t)|, \quad x \in C^{(1)}(0, T).$$

Then, for $x_0 \neq 0$,

$$F'(x_0, h) = \max_{t \in R} (h(t) \operatorname{sign} x_0(t)), \quad R = \{t \in [0, T]: |x_0(t)| = F(x_0)\},$$
$$K = \{h: h(t) \operatorname{sign} x_0(t) < 0, t \in R\},$$

and for $x_0 = 0$ we have $F'(x_0, h) = \|h\|$.

Example 7.7. Non-smooth integral functional:

$$F(x) = \int_0^T G(x(t), t)\, dt, \quad x \in C^{(n)}(0, T).$$

In contradistinction to Example 7.2, it is not assumed here that $G(x, t)$ is differentiable with respect to x or continuous in t. Let $G(x, t)$ satisfy a Lipschitz condition in x for arbitrary bounds on x (i. e., $|G(x_1, t) - G(x_2, t)| \leqslant L|x_1 - x_2|$ for $|x_1| \leqslant \rho$, $|x_2| \leqslant \rho$, $L = L(\rho)$); assume moreover that $G(x, t)$ is measurable in t for any x, and differentiable with respect to x in any direction for almost all $0 \leqslant t \leqslant T$ (i. e., we assume the existence of the limit

$$G_x(x, h, t) = \lim_{\varepsilon \to +0} \frac{G(x + \varepsilon h, t) - G(x, t)}{\varepsilon}\Big) .$$

Then $F(x)$ is differentiable at any point in any direction, and

$$F'(x_0, h) = \int_0^T G_x(x_0(t), h(t), t)\, dt.$$

Proof.

$$\frac{F(x_0 + \varepsilon h) - F(x_0)}{\varepsilon} = \int_0^T \varphi(\varepsilon, t)\, dt,$$

where

$$\varphi(\varepsilon, t) = \frac{G(x_0(t)+\varepsilon h(t), t) - G(x_0(t), t)}{\varepsilon}.$$

But, on the one hand,

$$\lim_{\varepsilon \to +0} \varphi(\varepsilon, t) = G_x(x_0(t), h(t), t)$$

for almost all $0 \leqslant t \leqslant T$, and, on the other,

$$| \varphi(\varepsilon, t) | \leqslant \frac{L\varepsilon \mid h(t) \mid}{\varepsilon} = L \mid h(t) \mid \leqslant L \| h \|$$

for all ε and t. We may therefore apply Lebesgue's theorem on limits under the integral sign (Natanson [1]), and this gives our formula for $F'(x_0, h)$.

Corollary. If $G(x, t)$ is measurable in t, differentiable with respect to x and $G_x(x, t)$ is bounded for bounded x, then

$$F'(x_0, h) = \int_0^T (G_x(x_0(t), t), h(t))dt.$$

FEASIBLE DIRECTIONS

We shall now describe how to determine the feasible directions. Throughout this lecture, K will denote the cone of feasible directions at a point x_0 for a set Q in a topological linear space.

We begin with two obvious remarks. First, the only interesting case is that in which x_0 is a boundary point of Q. For if $x_0 \in Q^0$, it follows directly from the definition that $K_b = E$ (any direction is feasible). Second, the set Q must have interior points. Otherwise the answer is trivial — the cone K_b is empty.

We consider the case in which the set Q is defined by some functional:

$$Q = \{x: F(x) \leqslant F(x_0)\}. \tag{8.1}$$

When F(x) is continuous, it follows from the above remarks that there is no point in studying the most general case $Q = \{x: F(x) \leqslant \lambda\}$, $\lambda \neq F(x_0)$. Let K_y denote the cone of directions of decrease at x_0 for the functional F(x).

Lemma 8.1. $K_y \subset K_b$.

Indeed, if $h \in K_y$, it follows from (6.1) that $F(x_0 + \varepsilon \bar{h}) \leqslant F(x_0) + \varepsilon\alpha < F(x_0)$ for all $\bar{h}$ in some neighborhood U of the vector h and all $0 < \varepsilon < \varepsilon_0$. Therefore, $x_0 + \varepsilon \bar{h} \in Q$ for all such $\bar{h}$ and ε, but this is precisely the definition of a feasible direction.

It turns out that in many important cases the cones K_y and K_b coincide.

Theorem 8.1. Let F(x) be differentiable at x_0 in any direction, $F'(x_0, h)$ convex in h, and assume that there exists $\tilde{h}$ such that $F'(x_0, \tilde{h}) < 0$. Then $K_b \subset \{h: F'(x_0, h) < 0\}$.

Proof. Let $h \in K_b$. Then $x_0 + \varepsilon h \in Q$ for $0 < \varepsilon < \varepsilon_0$, i.e., $F(x_0 + \varepsilon h) \leqslant F(x_0)$.

Hence it follows that $F'(x_0, h) \leqslant 0$. Furthermore, the cone K_b is open (Lemma 6.2). Hence there exists a neighborhood U of the vector h such that $U \subset K_b$. Take $\gamma > 0$ such that $h_\gamma = h - \gamma(h - \tilde{h}) \in U$. Then $F'(x_0, h_\gamma) \leqslant 0$ (since $h_\gamma \in K_b$), and it follows from the convexity of $F'(x_0, h)$ in h that

$$F'(x_0, h) = F'\left(x_0, \frac{1}{1+\gamma} h_\gamma + \frac{\gamma}{1+\gamma}\tilde{h}\right) \leqslant$$
$$\leqslant \frac{1}{1+\gamma} F'(x_0, h_\gamma) + \frac{\gamma}{1+\gamma} F'(x_0, \tilde{h}) \leqslant \frac{\gamma}{1+\gamma} F'(x_0, \tilde{h}) < 0.$$

Corollary. Assume that any one of the following conditions holds:

a) E is a Banach space, F(x) satisfies a Lipschitz condition in a neighborhood of the point x_0, F(x) is differentiable at x_0 in any direction, $F'(x_0, h)$ is convex in h, and there exists $\tilde{h}$ such that $F'(x_0, \tilde{h}) < 0$.

b) F(x) is a convex continuous functional and there exists $\tilde{x}$ such that $F(\tilde{x}) < F(x_0)$.

c) E is a Banach space, F(x) is differentiable at x_0, and $F'(x_0) \neq 0$.

Then

$$K_b = K_y = \{h: F'(x_0, h) < 0\}.$$

The proof follows directly from Theorem 8.1 and Lemma 8.1, using Theorems 7.3, 7.4, 7.5.

Using this theorem and the technique, considered in detail in the preceding lecture, for constructing the cone of directions of decrease, one easily finds the cones of feasible directions for various sets.

In conclusion we mention a special case of a set which is not defined by a functional, and nevertheless its cone of feasible directions has a simple description.

Theorem 8.2. Let Q be a convex set. Then the cone of feasible directions

K_b at the point x_0 is given by

$$K_b = \{\lambda(Q^0 - x_0),\ \lambda > 0\}$$
$$\text{(i. e.,}\quad K_b = \{h\colon h = \lambda(x - x_0),\ x \in Q^0,\ \lambda > 0\}).$$

The proof follows directly from the definition.

Lecture 9

TANGENT DIRECTIONS

The theorem of Lyusternik proved below is a powerful tool for the calculation of tangent directions. Before proceeding to a statement of the theorem, we recall the definition of a differentiable operator. Let E_1, E_2 be Banach spaces, $P(x)$ an operator (generally nonlinear) with domain in E_1 and range in E_2. Then $P(x)$ is said to be differentiable at a point $x_0 \in E_1$ if there exists a continuous linear operator A mapping E_1 into E_2, such that for all $h \in E_1$,

$$P(x_0+h)=P(x_0)+Ah+r(x_0, h),$$

where $\|r(x_0, h)\| = o(\|h\|)$. The operator A is called the (Fréchet-)derivative of the operator $P(x)$ and often denoted by $A = P'(x_0)$. It is clear that if $E_2 = R^1$ (i.e., $P(x)$ is a functional), this definition coincides with the previous definition (Lecture 7) of the derivative of a functional. The derivative of an operator possesses the usual properties of derivatives (rules for differentiation of sums, composite functions, etc.). The derivative of a continuous linear operator coincides with the operator.

Theorem 9.1 (Lyusternik). Let $P(x)$ be an operator mapping E_1 into E_2, differentiable in a neighborhood of a point x_0, $P(x_0) = 0$. Let $P'(x)$ be continuous in a neighborhood of x_0, and suppose that $P'(x_0)$ maps E_1 onto E_2 (i.e., the linear equation $P'(x_0)h = b$ has a solution h for any $b \in E_2$). Then the set of tangent directions K to the set $Q = \{x: P(x) = 0\}$ at the point x_0 is the subspace $K = \{h: P'(x_0)h = 0\}$.

The proof of this theorem (which is by no means trivial) may be found, e.g., in Lyusternik and Sobolev [1]. In the case $P'(x_0)E_1 \neq E_2$ one can assert only that

$K \subset \{h: P'(x_0)h = 0\}$ (this is quite easy to prove).

Using Lyusternik's theorem, we shall determine the cone of tangent directions (denoted throughout this lecture by K) for various sets Q at a point x_0.

Example 9.1. Let $x \in R^m$, $Q = \{x: G_i(x) = 0,\ i = 1, \ldots, n\}$, where $G_i(x)$ are functions continuously differentiable at the point x_0, $G_i(x_0) = 0$, $i = 1, \ldots, n$, and the vectors $G_i'(x_0)$, $i = 1, \ldots, n$, are linearly independent. Then $K = \{h \in R^n: (G_i'(x_0), h) = 0,\ i = 1, \ldots, n\}$.

In fact, here $E_1 = R^m$, $E_2 = R^n$, $P(x) = (G_1(x), \ldots, G_n(x))$, $P'(x_0)$ is the $m \times n$ matrix whose i-th column is the vector $G_i'(x_0)$. The fact that $P'(x_0)$ maps R^m onto R^n is equivalent to the linear independence of the columns of the matrix. Hence Lyusternik's theorem yields the desired result.

Note that the assumption that the vectors $G_i'(x_0)$, $i = 1, \ldots, n$, are independent is essential. For example, if $n = 1$, $G_1(x) = |x|^2$, $x_0 = 0$, then $G_1'(x_0) = 0$, and thus $\{h: (G_1'(x_0), h) = 0\} = R^m$, whereas in actual fact $K = \{0\}$.

Example 9.2. Consider the space $E_1 = C \times L_\infty$, i.e., the set of functions $x(t) \in C^{(n)}(0, T)$, $u(t) \in L_\infty^r(0, T)$, and the set

$$Q = \{x, u \in E_1: x(t) = c + \int_0^t \varphi(x(\tau), u(\tau), \tau)\, d\tau\}$$

(in other words, x and u satisfy the differential equation

$$\frac{dx(t)}{dt} = \varphi(x(t), u(t), t), \quad x(0) = c).$$

Let us determine the tangent subspace $K = \{\bar{x}(t), \bar{u}(t)\}$ of this constraint.

Introduce the operator

$$P(x, u) = x(t) - c - \int_0^t \varphi(x(\tau), u(\tau), \tau)\, d\tau$$

which maps E_1 into $C^{(n)}(0, T)$. Then $Q = \{x, u: P(x, u) = 0\}$. Assume that, for all bounded x, u and all $0 \leqslant t \leqslant T$, the derivatives $\varphi_x(x, u, t)$, $\varphi_u(x, u, t)$ exist, are continuous in x, u, measurable in t, and bounded (here $\varphi_x(x, u, t)$ and $\varphi_u(x, u, t)$ denote the matrices whose elements are the partial derivatives

$$\frac{\partial \varphi_i(x, u, t)}{\partial x_j}, \; i=1, \dots, n, \; j=1, \dots, n$$

and

$$\frac{\partial \varphi_i(x, u, t)}{\partial u_j}, \; i=1, \dots, n, \; j=1, \dots, r).$$

Then

$$P(x+\bar{x}, u+\bar{u}) - P(x, u) = \bar{x}(t) - \int_0^t [\varphi_x(x, u, \tau)\bar{x}(\tau) + \varphi_u(x, u, \tau)\bar{u}(\tau)]\, d\tau + \delta,$$

where δ is a remainder term for which an estimate is easily found:

$$\delta = o\left(\sqrt{\|\bar{x}\|_C^2 + \|\bar{u}\|_{L_\infty}^2}\right),$$

and the first term on the right-hand side is a linear operator of $\bar{x}$, $\bar{u}$, defined in C. Therefore, P(x, u) is differentiable, and

$$P'(x, u)(\bar{x}, \bar{u}) = \bar{x}(t) - \int_0^t [\varphi_x(x, u, \tau)\bar{x}(\tau) + \varphi_u(x, u, \tau)\bar{u}(\tau)]\, d\tau.$$

P'(x, u) is continuous in a neighborhood of x, u.

We now show that P'(x, u) maps E_1 onto C, i.e., the equation

$$\bar{x}(t) - \int_0^t [\varphi_x(x, u, \tau)\bar{x}(\tau) + \varphi_u(x, u, \tau)\bar{u}(\tau)]\, d\tau = a(t)$$

has a solution $\bar{x}, \bar{u}$ for any $a(t) \in C^{(n)}(0, T)$. Set $\bar{u}(t) \equiv 0$; then the equation becomes

$$\bar{x}(t)=a(t)+\int_0^t \varphi_x(x, u, \tau)\bar{x}(\tau)\, d\tau.$$

This integral equation is a linear Volterra equation of the second kind, and the theory of integral equations (see Kantorovich and Akilov [1]) tells us that it has a solution $\bar{x}(t) \in C$ for any $a(t) \in C$.

Thus, all the assumptions of Lyusternik's theorem hold for the operator $P(x, u)$, and therefore the tangent subspace K consists of all pairs $\bar{x}(t)$, $\bar{u}(t)$ which satisfy the integral equation

$$\bar{x}(t)=\int_0^t [\varphi_x(x, u, \tau)\bar{x}(\tau)+\varphi_u(x, u, \tau)\bar{u}(\tau)]\, d\tau,$$

or, equivalently, the linear differential equation

$$\frac{d\bar{x}(t)}{dt} = \varphi_x(x, u, t)\bar{x}(t)+\varphi_u(x, u, t)\bar{u}(t), \quad \bar{x}(0)=0;$$

this equation is usually known as the variational equation.

Example 9.3. Let us slightly complicate the preceding example, by adding the constraint $x(T) = d$, $d \in R^n$, so that

$$Q=\{x, u \in E_1\colon x(t)=c+\int_0^t \varphi(x(\tau), u(\tau), \tau)d\tau,\ 0 \leqslant t \leqslant T,\ x(T)=d\}.$$

Let us determine the tangent subspace to Q.

Introduce the operator

$$P(x, u)=(x(t)-c-\int_0^t \varphi(x, u, \tau)\, d\tau,\ x(T)-d)$$

which maps E_1 into $E_2 = C^{(n)}(0, T) \times R^n$ (i.e., it maps each pair $x(t) \in C$, $u(t) \in L_\infty$ onto the pair

$$x(t)-c-\int_0^t \varphi(x, u, \tau)\, d\tau \in C,\ x(T)-d \in R^n).$$

Using the same assumptions and arguments as in the preceding example, we see that P(x) is continuously differentiable, and

$$P'(x, u)(\bar{x}, \bar{u}) = \left(\bar{x}(t) - \int_0^t [\varphi_x(x,u,\tau)\, \bar{x}(\tau) + \varphi_u(x, u, \tau)\, \bar{u}(\tau)]d\tau, \bar{x}(T)\right).$$

We wish to determine when this linear operator maps E_1 onto E_2. We introduce a definition:

Nondegeneracy condition 9.1. Let $D \subset R^n$ be the set of all vectors $\bar{x}(T)$, where $\bar{x}(t)$ satisfies the differential equation

$$\frac{d\bar{x}(t)}{dt} = A(t)\,\bar{x}(t) + B(t)\,\bar{u}(t), \quad \bar{x}(0) = 0, \tag{9.1}$$

with $\bar{u}(t)$ ranging over the entire space $L_\infty^{(r)}(0, T)$. Here A(t) and B(t) are $n \times n$ and $n \times r$ matrices, $\bar{u}(t) \in R^r$, $\bar{x}(t) \in R^n$. This equation is said to be nondegenerate* if $D = R^n$.

We claim that if the nondegeneracy condition holds for the system (9.1) with $A(t) = \varphi_x(x(t), u(t), t)$, $B(t) = \varphi_u(x(t), u(t), t)$, then $P'(x, u)$ maps E_1 onto E_2. Let $(a(t), b) \in E_2$. Take $z(t) \in C^{(n)}(0, T)$ such that

$$z(t) = a(t) + \int_0^t \varphi_x(x, u, \tau)\, z(t)\, d\tau, \quad 0 \leqslant t \leqslant T$$

(it was shown in Example 9.2 that this can be done).

Now, since the system is nondegenerate, we can find

$$\bar{u}(t) \in L_\infty^{(r)}(0, T), \; y(t) \in C^n(0, T),$$

such that

$$\frac{dy}{dt} = \varphi_x(x, u, t) y(t) + \varphi_u(x, u, t)\, \bar{u}(t), \; y(0) = 0, \; y(T) = b - z(T).$$

---

* A system of differential equations satisfying this condition is sometimes said to be completely controllable.

Finally, set $\overline{x}(t) = y(t) + z(t)$. Then

$$P'(x, u)(\overline{x}, \overline{u}) = (\overline{x}(t) - \int_0^t [\varphi_x(x, u, \tau)\overline{x}(\tau) + \varphi_u(x, u, \tau)\overline{u}(\tau)]d\tau, \overline{x}(T)$$

$$= (z(t) + y(t) - \int_0^t \varphi_x(x, u, \tau)\, z(\tau)\, d\tau - y(t), b) = (a(t), b),$$

as required.

Applying Lyusternik's theorem, we arrive at the final result:

Under the assumptions of Example 9.2, let the system

$$\frac{d\overline{x}}{dt} = \varphi_x(x, u, t)\overline{x}(t) + \varphi_u(x, u, t)\,\overline{u}(t) \qquad , \quad \overline{x}(0) = 0$$

be nondegenerate. Then

$$K = \{\overline{x}, \overline{u}: \overline{x}(t) = \int_0^t [\varphi_x(x, u, \tau)\overline{x}(\tau) + \varphi_u(x, u, \tau)\overline{u}(\tau)]\, d\tau,$$
$$\overline{x}(T) = 0\}.$$

Alternatively, in terms of differential equations,

$$K = \left\{\overline{x}, \overline{u}: \frac{d\overline{x}(t)}{dt} = \varphi_x(x, u, t)\,\overline{x}(t) + \varphi_u(x, u, t)\,\overline{u}(t), \right.$$
$$\left. \overline{x}(0) = \overline{x}(T) = 0 \right\}.$$

The nondegeneracy condition 9.1 is not easy to verify. Let us try to formulate it in a more explicit way.

N o n d e g e n e r a c y   c o n d i t i o n 9.2. The system (9.1) is said to be nondegenerate if any nonzero solution $\Psi(t)$ of the equation

$$\frac{d\psi(t)}{dt} = -A^*(t)\,\psi(t), \qquad (9.2)$$

satisfies the condition $B^*(t)\Psi(t) \not\equiv 0$ (more precisely, $B^*(t)\Psi(t)$ is nonzero on a set of positive measure).

L e m m a 9.1. The nondegeneracy condition 9.2 implies condition 9.1.

P r o o f. Suppose that condition 9.1 does not hold, i.e., $D \neq R^n$. Since D is a subspace (this is an obvious consequence of the linearity of the equation), it follows that there exists a nonzero vector $a \in R^n$ orthogonal to D (i.e., $(a, \bar{x}(T)) = 0$ for any $\bar{x}(T)$), which is a solution of equation (9.1).

Consider the system

$$\frac{d\psi(t)}{dt} = -A^*(t)\psi(t), \quad \psi(T) = a.$$

Since $a \neq 0$, it follows that also $\Psi(t) \not\equiv 0$. Then, for any solution $\bar{x}(t)$ of (9.1),

$$0 = \int_0^T \left(\frac{d\psi(t)}{dt} + A^*(t)\psi(t), \bar{x}(t)\right) dt =$$

$$= (\psi(T), \bar{x}(T)) - (\psi(0), \bar{x}(0)) - \int_0^T \left(\psi(t), \frac{d\bar{x}(t)}{dt}\right) dt +$$

$$+ \int_0^T (\psi(t), A(t)\bar{x}(t))\, dt = (a, \bar{x}(T)) - \int_0^T (\psi(t), B(t)\bar{u}(t))\, dt =$$

$$= -\int_0^T (B^*(t)\psi(t), \bar{u}(t))\, dt,$$

i.e., for any $\bar{u}(t) \in L_\infty$ we have

$$\int_0^T (B^*(t)\psi(t), \bar{u}(t))\, dt = 0.$$

But this is possible only if $B^*(t)\Psi(t) = 0$ almost everywhere, which contradicts condition 9.2.

Thus, the equation of the tangent subspace derived above is also valid when the nondegeneracy condition is given in the formulation 9.2. Now condition 9.2 is quite constructive. In fact, it is sufficient to find a fundamental system of solutions of the equation

$$\frac{d\psi(t)}{dt} = -A^*(t)\psi(t)$$

(i. e., to solve the matrix equation

$$\frac{d\Psi(t)}{dt} = -A^*(t)\Psi(t),\ \Psi(0) = I,$$

where I is the unit matrix, $\Psi(t)$ an $n \times n$ matrix), and then to find the matrix $B^*(t)\Psi(t)$. If the columns $\xi_i(t)$ of this matrix are linearly independent as functions of t, i. e., $\sum_{i=1}^{n} \lambda_i \xi_i(t) \neq 0$ for $\lambda = (\lambda_1, \ldots, \lambda_n) \neq 0$, then the system is nondegenerate. Indeed, any nonzero solution $\psi(t)$ of the system (9. 2) has the form $\Psi(t)\lambda$, $\lambda \neq 0$, and therefore $B^*(t)\psi(t) = B^*(t)\Psi(t)\lambda = \sum_{i=1}^{n} \lambda_i \xi_i(t)$.

Condition 9. 2 is considerably simplified when the matrices A and B are constant, but we shall not discuss this here.

CALCULATION OF DUAL CONES

In the preceding lectures we learned how to find the cone of directions of decrease, the cone of feasible directions, and the tangent subspace. For application of the Dubovitskii-Milyutin theorem to determine necessary conditions for an extremum, it remains to show how one constructs dual cones. This we now proceed to do. Some results in this connection were presented in Lecture 5 (Lemmas 5.2 to 5.10).

We begin with a few simple calculations of dual cones (primarily for subspaces).

Theorem 10.1. Let K be a subspace. Then $K^* = \{f \in E': f(x) = 0$ for all $x \in K\}$ (this set is sometimes known as the annihilator of K).

Indeed, let $f(x) > 0$ for some $x \in K$, $f \in K^*$. Since $-x \in K$ and $f(-x) = -f(x) < 0$, this contradicts the fact that $f \in K^*$.

Theorem 10.2. Let

$$f \in E', \ K_1 = \{x: f(x) = 0\}, \ K_2 = \{x: f(x) \geqslant 0\}, \ K_3 = \{x: f(x) > 0\}.$$

Then $K_1^* = \{\lambda f, -\infty < \lambda < \infty$, $K_2^* = \{\lambda f, 0 \leqslant \lambda < \infty$, $K_3^* = E'$ for $f = 0$ and $K_3^* = K_2^*$ for $f \neq 0$.

In fact, by Theorem 10.1, if $g \in K_1^*$, then $g(x) = 0$ on K_1. Hence, by Lemma 2.3, $g = \lambda f$. Since $K_1 \subset K_2$, it follows (Lemma 5.5) that $K_2^* \subset K_1^*$. But if $g = \lambda f \in K_2^*$, then $g(x) = \lambda f(x) \geqslant 0$ for all $x \in K_2$, so that $\lambda \geqslant 0$. Finally, if $f = 0$ then $K_3 = \emptyset$, and hence $K_3^* = E'$; if $f \neq 0$, then $\overline{K}_3 = K_2$, so that $K_3^* = K_2^*$ (Lemma 5.3).

Theorem 10.3. Let $E = E_1 \times E_2$, where E_1 and E_2 are topological linear spaces; let A be a linear operator mapping E_1 into E_2, and set

$$K = \{x \in E,\ x = (x_1, x_2):\ Ax_1 = x_2\}.$$

Then

$$K^* = \{f \in E',\ f = (f_1, f_2):\ f_1 = -A^* f_2\}.$$

Proof. If $f \in K^*$, then for all $x \in K$ we have, by Theorem 10.1,

$$0 = (f, x) = (f_1, x_1) + (f_2, x_2) = (f_1 + A^* f_2, x_1),$$

whence it follows that $f_1 + A^* f_2 = 0$. Conversely, if $f = (-A^* f_2, f_2)$, then, for $x \in K$,

$$(f, x) = (-A^* f_2, x_1) + (f_2, x_2) = -(f_2, Ax_1 - x_2) = 0.$$

We can now prove the important theorem of Minkowski and Farkas, which specifies the form of the dual cone for cones of a very general type. We shall use the same notation as in Theorem 10.3.

Theorem 10.4 (Minkowski-Farkas). Let K_2 be a convex cone in E_2 with apex at 0, $K_1 = \{x_1 \in E_1:\ Ax_1 \in K_2\}$, and assume that either of the following two conditions holds:

a) There exists $\tilde{x}_1 \in E_1$ such that $A\tilde{x}_1 \in K_2^0$.

b) E_1, E_2 are finite-dimensional, K_2 is the positive octant of E_2 (i.e.,

$$K_1 = \{x \in R^m:\ (a^i, x) \geqslant 0,\ a^i \in R^m,\ i = 1, \dots, n\}$$

or, in matrix notation, $K_1 = \{x \in R^m:\ Ax \geqslant 0\}$, where A is an $n \times m$ matrix).

Then $K_1^* = A^* K_2^*$ (i.e., in case b), $K_1^* = \sum a^i y_i,\ y_i \geqslant 0,\ i = 1, \dots, n$, or $K_1^* = \{A^* y,\ y \geqslant 0\}$).

Proof. Case a). Denote $K = \{x \in E:\ x = (x_1, x_2),\ x_2 \in K_2\}$,

$L = \{x \in E: Ax_1 = x_2\}$. Then K is a convex cone with apex at 0 and L is a subspace such that $K^0 \cap L \neq \emptyset$. By Lemma 5.9, $(K \cap L)^* = K^* + L^*$. Now let $\tilde{f}_1 \in K_1^*$. Then $\tilde{f} = (\tilde{f}_1, 0) \in (K \cap L)^*$ (since if $x \in K \cap L$, then $x = (x_1, Ax_1)$, $x_1 \in K_1$, and therefore $\tilde{f}(x) = \tilde{f}_1(x_1) \leqslant 0$).

Consequently, by what we have proved, $\tilde{f} \in K^* + L^*$. But $K^* = \{f = (f_1, f_2): f_1 = 0, f_2 \in K_2^*\}$ (this is obvious), $L^* = \{f = (f_1, f_2): f_1 = -A^* f_2\}$ (by Theorem 10.3). Hence there exists $\tilde{f}_2 \in K_2^*$ such that $\tilde{f}_1 = A^* \tilde{f}_2$, that is, $K_1^* \subset A^* K_2^*$. The inverse inclusion is obvious: if $f_1 = A^* f_2$, $f_2 \in K_2^*$, then

$$(f_1, x_1) = (A^* f_2, x_1) = (f_2, Ax_1) \geqslant 0$$

for $x \in K_1$ (i.e., for $Ax_1 \in K_2$).

C a s e b). Set $Q_i = \{x: (a^i, x) \geqslant 0\}$; then $K_1 = \bigcap_{i=1}^{n} Q_i$. By Theorem 10.2, $Q_i^* = \{a^i y_i: y_i \geqslant 0\}$. Consider

$$\sum_{i=1}^{n} Q_i^* = \left\{ \sum_{i=1}^{n} a^i y_i,\ y_i \geqslant 0,\ i = 1, \dots, n \right\}.$$

It can be shown that any set of the type

$$\left\{ \sum_{i=1}^{n} a^i y_i,\ y_i \geqslant 0,\ i = 1, \dots, n \right\}$$

is closed in R^m (see, e.g., Yudin and Gol'shtein [1], p. 237). Since all topologies of R^m coincide, the set is also weakly* closed in R^m (see Lecture 2). Hence, by the Corollary to Lemma 5.8,

$$\left(\bigcap_{i=1}^{n} Q_i\right)^* = \sum_{i=1}^{n} Q_i^*, \text{ i.e.,}$$

$$K_1^* = \left\{ \sum_{i=1}^{n} a^i y_i,\ y_i \geqslant 0,\ i = 1, \dots, n \right\}.$$

It can be shown that conditions a) and b) are special cases of the following condition: <u>The cone $A^*K_2^*$ is weakly* closed in E_1^*</u>. This also implies the equality $K_1^* = A^*K_2^*$.

For the moment, we shall limit ourselves to one application of Theorem 10.4.

Example 10.1.

$$K = \{x \in R^m\colon (a^i, x) \geqslant 0,\ i = 1, \dots, k,\ (a^i, x) = 0,\\ i = k+1, \dots, n\},$$

where $a^i \in R^m$, $i = 1, \dots, n$.

Then

$$K^* = \left\{f \in R^m\colon f = \sum_{i=1}^{n} \lambda_i a^i,\ \lambda_i \geqslant 0,\ i = 1, \dots, k\right\},$$

which follows directly from Case b) of Theorem 10.4 (the constraints $(a^i, x) = 0$, $i = k+1, \dots, n$ must first be rewritten as inequalities $(a^i, x) \geqslant 0$, $(-a^i, x) \geqslant 0$, $i = k+1, \dots, n$). In different notation: if $K = \{x\colon A_1 x \geqslant 0,\ A_2 x = 0\}$, then $K^* = A_1^* y_1 + A_2^* y_2,\ y_1 \geqslant 0$. Here $x \in R^m$, $y_1 \in R^k$, $y_2 \in R^{n-k}$; A_1 and A_2 are $k \times m$ and $(n-k) \times m$ matrices, respectively, and A_1^*, A_2^* are their transposes. All inequalities refer to coordinates.

We present a few more important examples of dual cones.

Example 10.2. $E = C^{(n)}(0, T)$, R is a closed set in $[0, T]$, $K = \{x \in C\colon x(t) \geqslant 0 \text{ for } t \in R\}$. Then, for any $f \in K^*$, there exists a nonnegative vector measure $d\mu(t)$ with support R, such that

$$f(x) = \int_0^T (x(t), d\mu(t)) = \int_R (x(t), d\mu(t)).$$

This assertion is simply a rephrased version of Riesz's theorem on the general form of a nonnegative linear functional in C (see, e.g., Kantorovich and

Akilov [1]).

Example 10.3. $E = C^{(n)}(0, T)$, R is a closed subset of [0, T],

$$h(t) \in C(0, T),\ h(t) \neq 0 \text{ on } R,$$

$$K = \{x \in C(0, T):\ (h(t), x(t)) \geqslant 0 \text{ for } t \in R\}.$$

Then, for any $f \in K^*$, there exists a nonnegative scalar measure $d\mu(t)$ with support R, such that

$$f(x) = \int_0^T (h(t), x(t))\, d\mu(t) = \int_R (h(t), x(t))\, d\mu(t).$$

Indeed, take $E_1 = C^{(n)}(0, T)$, $E_2 = C^{(1)}(0, T)$, A the linear operator mapping E_1 into E_2 defined as scalar multiplication by h(t): $Ax = (x(t), h(t))$, and consider the cone

$$K_2 = \{y(t) \in C^{(1)}(0, T):\ y(t) \geqslant 0 \text{ for } t \in R\}.$$

Then $K_2 = AK$, where K is the original cone. We use Theorem 10.4. If we take $\tilde{x}(t) = h(t)$, then $A\tilde{x} = |h(t)|^2 > 0$ on R, i.e., $A\tilde{x} \in \overset{0}{K}_2$, and condition a) holds. By Example 10.2, K_2^* is determined by a nonnegative measure $d\mu(t)$ on R. Therefore, by Theorem 10.4, for any $f \in K^*$ there exists $f_2 \in K_2^*$ such that $f = A^* f_2$, i.e.,

$$(f, x) = (A^* f_2, x) = (f_2, Ax) = \int_0^T (h(t), x(t))\, d\mu(t).$$

Let Q be a set in a topological linear space E, $x_0 \in Q$, K_b the cone of feasible directions for Q at x_0 and K_k the cone of tangent directions for Q at x_0. It turns out that in some cases the duals of these cones coincide with the set of linear functionals which are <u>supports</u> for Q at the point x_0 (see Lecture 4). Denote the latter set by Q^*, i.e., $Q^* = \{f \in E': f(x) \geqslant f(x_0) \text{ for all } x \in Q\}$.

Theorem 10.5. Let Q be a closed convex set. Then $K_k^* = Q^*$. If moreover

$Q^0 \neq \emptyset$, then $K_b^* = Q^*$.

Proof. Let $f \in Q^*$, $h \in K_k$. Given $0 < \varepsilon < \varepsilon_0$, there exists $x(\varepsilon) \in Q$, $x(\varepsilon) = x_0 + \varepsilon h + r(\varepsilon)$, such that $\frac{1}{\varepsilon} r(\varepsilon) \in U$ for any neighborhood U of zero and sufficiently small ε (see Lecture 6). Since $x(\varepsilon) \in Q$ and $f \in Q^*$, it follows that $(f, x(\varepsilon)) \geqslant (f, x_0)$, and therefore

$$(f, h) = \frac{(f, x(\varepsilon)) - (f, x_0)}{\varepsilon} - \frac{(f, r(\varepsilon))}{\varepsilon} \geqslant -\left(f, \frac{r(\varepsilon)}{\varepsilon}\right).$$

But $(f, \frac{r(\varepsilon)}{\varepsilon}) \longrightarrow 0$ as $\varepsilon \longrightarrow +0$ (for, since f is continuous, for every $\delta > 0$ there exists a neighborhood U of zero such that $|(f, x)| \leqslant \delta$ for $x \in U$, and therefore $\left|(f, \frac{r(\varepsilon)}{\varepsilon})\right| < \delta$ for sufficiently small $\varepsilon > 0$), and therefore $(f, h) \geqslant 0$, i. e., $f \in K_k^*$.

Conversely, let $f \in K_k^*$, $x \in Q$. Then $h = x - x_0$ is a tangent direction (for $x_0 + \varepsilon h = (1 - \varepsilon)x_0 + \varepsilon x \in Q$ when $0 < \varepsilon < 1$, since Q is convex). Since $f \in K_k^*$, it follows that $(f, h) \geqslant 0$, i. e., $(f, x) \geqslant (f, x_0)$, and this means that $f \in Q^*$.

We have thus proved that $Q^* = K_k^*$.

If $Q^0 \neq \emptyset$, then, by Theorem 8.2,

$$K_b = \{h \colon h = \lambda (x - x_0),\ \lambda \geqslant 0,\ x \in Q^0\}.$$

Therefore, if $f \in K_b^*$, then

$$(f, \lambda (x - x_0)) \geqslant 0, \quad \lambda (f, x) \geqslant \lambda (f, x_0)$$

for all $x \in Q^0$ and all $\lambda \geqslant 0$, in particular, for $\lambda = 1$. But $\overline{(Q^0)} = \overline{Q}$ when $Q^0 \neq \emptyset$ (Lemma 2.5), and hence, by the continuity of (f, x), the fact that $(f, x) \geqslant (f, x_0)$ for all $x \in Q^0$ implies that $(f, x) \geqslant (f, x_0)$ for all $x \in \overline{Q}$, i. e., $f \in Q^*$ and $K_b^* \subset K_k^* = Q^*$. On the other hand, $K_b \subset K_k$, and therefore (Lemma 5.5) $K_b^* \supset K_k^* = Q^*$. Thus $K_b^* = Q^*$.

Thus, we see that in many cases determination of dual cones is equivalent to determination of the supporting functionals. We consider two important examples.

Example 10.4. Let Q be a polyhedron in R^m:

$$Q = \{x: (a^i, x) \geq b_i, \quad i = 1, \ldots, n\}, \ a^i \in R^m, \ b_i \in R^1, \ i = 1, \ldots, n$$

Then

$$K_h{}^* = Q^* = \left\{ \sum_{i=1}^{n} \lambda_i a^i, \ \lambda_i \geq 0, \ \lambda_i [(a^i, x_0) - b_i] = 0 \right\}.$$

Proof. Let

$$f = \sum_{i=1}^{n} \lambda_i a^i, \ \lambda_i \geq 0, \ \lambda_i [(a^i, x_0) - b_i] = 0.$$

Then for any $x \in Q$,

$$(f, x) = \sum_{i=1}^{n} \lambda_i (a^i, x) \geq \sum_{i=1}^{n} \lambda_i b_i = \sum_{i=1}^{n} \lambda_i (a^i, x_0) = (f, x_0), \text{ i. e. } f \in Q^*.$$

We now prove that $K_k \supset \{h: (a^i, h) \geq 0, \ i \in I\}$, where $I = \{i: (a^i, x_0) = b_i\}$. Indeed, if $(a^i, h) \geq 0$, $i \in I$, take

$$\varepsilon_0 = \min_{i \in I} \frac{(a^i, x_0) - b_i}{\| a^i \| \ \| h \|} > 0,$$

then $x_0 + \varepsilon h \in Q$ for $0 < \varepsilon < \varepsilon_0$ (since

$$(a^i, x_0 + \varepsilon h) = (a^i, x_0) + \varepsilon (a^i, h) \geq (a^i, x_0) = b_i$$

for $i \in I$ and

$$(a^i, x_0 + \varepsilon h) = (a^i, x_0) + \varepsilon (a^i, h) \geq (a_i, x_0) - \varepsilon \| a^i \| \ \| h \| \geq b_i$$

for $i \notin I$). Therefore $h \in K_k$. Hence it follows that $K_k^* \subset \{h: (a^i, h) \geq 0, \ i \in I\}^*$

(Lemma 5.5). But

$$\{h: (a^i, h) \geqslant 0, \ i \in I\}^* = \{\Sigma \lambda_i a^i, \ \lambda_i \geqslant 0, \ i \in I\}$$

(Example 10.1). Thus $Q^* \supset \{\sum \lambda_i a^i, \ \lambda_i \geqslant 0, \ i \in I\} \supset K_k^*$, but by Theorem 10.5 $K_k^* = Q^*$, and this completes the proof.

In matrix notation, the result just proved is:

$$\text{if } Q = \{x: Ax \geqslant b \ , \text{ then}$$

$$Q^* = K_k^* = \{A^*y, \ y \geqslant 0, \ y_i [Ax_0 - b]_i = 0, \ i = 1, \ldots, n\} .$$

We now consider another important example. Let $Q = \{x \in L_\infty^{(r)}(0, T): x(t) \in M$ for almost all $0 \leqslant t \leqslant T\}$, where M is a subset of R^r. It is quite difficult to determine the set of supporting functionals for this set at a point (suffice it to say that the structure of the space L_∞^* is comparatively complex). Nevertheless, we can give a description of the general form of supporting functionals defined by integrals.

Example 10.5. Let $Q = \{x \in L_\infty^{(r)}(0, T): x(t) \in M$ for almost all $0 \leqslant t \leqslant T\}$, $M \subset R^r$, $x_0 \in Q$. Then, if the linear functional defined by

$$f(x) = \int_0^T (a(t), \ x(t))dt, \ a(t) \in L_1^{(r)}(0, T),$$

is a support to Q at the point x_0, then $(a(t), x - x_0(t)) \geqslant 0$ for all $x \in M$ and almost all $0 \leqslant t \leqslant T$ (i.e., for almost all $0 \leqslant t \leqslant T$ the vector $a(t) \in R^r$ is a support to M at the point $x_0(t)$).

Proof. Suppose the assertion false. That is to say, there exists a subset $R_1 \subset [0, T]$, $\mu(R_1) \neq 0$ (where $\mu(R)$ denotes the Lebesgue measure of R), such that for every $t \in R_1$ there exists $\tilde{x}(t) \in M$ with $(a(t), \ \tilde{x}(t) - x_0(t)) < 0$. By Lusin's theorem (Natanson [1]) there exist subsets $R_2 \subset [0, T]$, $\mu(R_2) < \varepsilon/2$, $R_3 \subset [0, T]$,

$\mu(R_3) < \varepsilon/2$, such that a(t) is continuous on $[0, T] - R_2$, $x_0(t)$ is continuous on $[0, T] - R_3$. Since $\mu(r_2) + \mu(R_3) < \mu(R_1)$, there exists a point $t_0 \in R_1$, $t_0 \notin R_2 \cup R_3$. Now, since a(t) and $x_0(t)$ are continuous at t_0 and $(a(t_0), \tilde{x}(t_0) - x_0(t_0)) = \gamma < 0$, there exists $R_4 \subset [0, T]$, $\mu(R_4) > 0$, such that $(a(t), \tilde{x}(t_0) - x_0(t)) \leqslant \gamma/2$ for $t \in R_4$. Consider the function

$$x_1(t) = \begin{cases} \tilde{x}(t_0) & \text{if } t \in R_4, \\ x_0(t) & \text{if } t \in [0, T] - R_4. \end{cases}$$

Then $x_1 \in Q$. At the same time,

$$f(x_1) = \int_0^T (a(t), x_0(t))\, dt + \int_{R_4} (a(t), \tilde{x}(t_0) - x_0(t))\, dt \leqslant$$
$$\leqslant f(x_0) + \frac{\gamma}{2} \mu(R_4),$$

i.e., $f(x_1) < f(x_0)$, contradicting the fact that f is a support to Q at the point x_0.

Lecture 11

LAGRANGE MULTIPLIERS AND THE KUHN-TUCKER THEOREM

Using the technique described above, we shall now proceed to derive necessary conditions for an extremum in the various problems listed in the first lecture.

11.1. Conditional extremum (equality constraints):

$$\min F_0(x), \quad F_i(x) = 0, \; i = 1,\dots,n. \tag{11.1}$$

Here $x \in E$, where E is a Banach space; $F_0(x), \dots, F_n(x)$ are functionals on E.

Theorem 11.1. Let x_0 be a solution of the problem (11.1), and assume that the functionals $F_i(x)$, $i = 0, \dots, n$ are continuously differentiable in a neighborhood of x_0. Then there exist numbers $\lambda_0, \dots, \lambda_n$, not all zero, $\lambda_0 \geq 0$, such that

$$\lambda_0 F_0'(x_0) + \lambda_1 F_1'(x_0) + \dots + \lambda_n F_n'(x_0) = 0. \tag{11.2}$$

Proof. If $F_i'(x_0)$, $i = 1, \dots, n$, are linearly dependent:

$$\sum_{i=1}^{n} \gamma_i F_i'(x_0) = 0, \quad \sum_{i=1}^{n} \gamma_i^2 \neq 0,$$

then we can take $\lambda_0 = 0$, $\lambda_i = \gamma_i$, $i = 1, \dots, n$, and (11.2) will hold. It will therefore suffice to consider the case in which $F_i'(x_0)$, $i = 1, \dots, n$, are linearly independent. In this case, the tangent subspace K_1 to the set $Q = \{x: F_i(x) = 0, \; i = 1, \dots, n\}$ was described in Example 9.1: $K_1 = \{h: (F_i'(x), h) = 0, \; i = 1, \dots, n\}$. The dual cone was determined in Example 10.1: $K_1^* = \left\{ \sum_{i=1}^{n} \lambda_i F_i'(x_0) \right\}$. The cone K_0 of directions of decrease for a differentiable functional is given, according to Theorem 7.5, by $K_0 = \{h: (F_0'(x_0), h) < 0\}$, and the dual cone is $K_0^* = \{\lambda_0 F_0'(x_0), \; 0 \leq \lambda_0\}$

provided $F_0'(x_0) \neq 0$, by Theorem 10.2 (the case $F_0'(x_0) = 0$ is trivial, since we can then take $\lambda_0 = 1$, $\lambda_i = 0$, $i = 1, \ldots, n$). Application of the Dubovitskii-Milyutin theorem now yields (11.2).

Note that if $F_i'(x_0)$, $i = 1, \ldots, n$, are linearly independent, then, as follows from the proof of the theorem, $\lambda_0 \neq 0$ and (11.2) may be rewritten as

$$F_0'(x_0) + \lambda_1 F_1'(x_0) + \ldots + \lambda_n F_n'(x_0) = 0. \tag{11.3}$$

In particular, if there are no constraints we get the usual condition for an unconditional extremum, $F_0'(x_0) = 0$. The numbers $\lambda_1, \ldots, \lambda_n$ are usually called Lagrange multipliers. In the finite-dimensional case, condition (11.3) is well known from elementary analysis (rule of Lagrange multipliers). The general form of the rule was obtained by Lyusternik in 1934. Note that condition (11.2) is completely symmetric with respect to all the functionals, both that being minimized and those appearing in the constraints (reciprocity principle).

11.2. The linear programming problem (inequality constraints):

$$\begin{gathered} \min F_0(x), \\ F_i(x) \leqslant 0,\ i = 1, \ldots, n. \end{gathered} \tag{11.4}$$

Here $F_i(x)$, $i = 1, \ldots, n$, have the same meaning as in problem 11.1.

Theorem 11.2. Let x_0 be a solution of the problem (11.4), and assume that the functionals $F_i(x)$, $i = 0, \ldots, n$, are differentiable at x_0. Then there exist numbers λ_i, $i = 0, \ldots, n$, not all zero, such that

$$\begin{gathered} \lambda_i \geqslant 0,\ i = 0, \ldots, n,\ \lambda_i F_i(x_0) = 0,\ i = 1, \ldots, n, \\ \lambda_0 F_0'(x_0) + \lambda_1 F_1'(x_0) + \ldots + \lambda_n F_n'(x_0) = 0. \end{gathered} \tag{11.5}$$

Proof. Set $Q_i = \{x: F_i(x) \leqslant 0\}$, $i = 1, \ldots, n$, and denote the cone of feasible directions for Q_i at x_0 by K_i. If $F_i(x_0) < 0$, we have $K_i = E$; while if $F_i(x_0) = 0$ and

$F_i'(x_0) \neq 0$, then $K_i = \{h: (F_i'(x_0), h) < 0\}$ (Corollary to Theorem 8.1). Finally, the case $F_i(x_0) = 0$, $F_i'(x_0) = 0$ is trivial, since we can then take $\lambda_i = 1$, $\lambda_j = 0$ for $j \neq i$, and condition (11.5) will hold. It follows from Theorem 10.2 that $K_i^* = 0$ if $F_i(x_0) < 0$ and $K_i^* = \{\lambda_i F_i'(x_0),\ 0 \leq \lambda_i\}$ if $F_i(x_0) = 0$, $F_i'(x_0) \neq 0$. The cone K_0 of directions of decrease for $F_0(x)$ at x_0 is

$$K_0 = \{h:\ (F_0'(x_0),\ h) < 0\},\quad K_0{}^* = \{\lambda_0 F_0'(x_0),\quad 0 \leqslant \lambda_0\}$$

for $F_0'(x_0) \neq 0$. Hence, using Theorem 6.1, we get (11.5).

We shall now improve this result slightly, indicating when it can be assumed that $\lambda_0 \neq 0$ (problems of this type are known as <u>nondegenerate</u>).

Theorem 11.3. Under the assumptions of Theorem 11.2, assume that one of the following conditions holds:

a) The functionals $F_i(x)$, $i = 1, \ldots, n$, are convex, and there exists $\tilde{x}$ such that $F_i(\tilde{x}) < 0$, $i = 1, \ldots, n$.

b) The constraints are linear:

$$F_i(x) = (a^i,\ x) - b_i,\quad a^i \in E',\quad b_i \in R^1,\quad i = 1, \ldots, n.$$

Then $\lambda_0 \neq 0$, i.e., the condition for a minimum at x_0 is:

$$\begin{gathered}\lambda_i \geqslant 0,\quad i = 1, \ldots, n,\quad \lambda_i F_i(x_0) = 0,\quad i = 1, \ldots, n,\\ F_0'(x_0) + \lambda_1 F_1'(x_0) + \ldots + \lambda_n F_n'(x_0) = 0.\end{gathered} \tag{11.6}$$

Proof. In Case a), we can use Remark 3 to Theorem 6.1, since $\tilde{x} - x_0$ is a feasible direction. In Case b), we regard $Q = \{x:\ (a^i, x) \leq b_i,\ i = 1, \ldots, n\}$ as one constraint (possibly containing no interior points). The dual cone of the cone of tangent directions to Q at x_0 is

$$K_1{}^* = \left\{ \sum_{i=1}^{n} \lambda_i a^i,\ \lambda_i \geqslant 0,\ \lambda_i\left((a^i,\ x_0) - b_i\right) = 0,\ i = 1, \ldots, n \right\}$$

(Example 10.4). Using Theorem 6.1, we now get the desired result.

Let us apply Theorem 11.3 to the linear programming problem

$$\min (c, x), \quad Ax \geqslant b, \tag{11.7}$$

where $x \in R^m$, A is an $n \times m$ matrix, and $b \in R^n$. We get the following necessary condition for a minimum at the point x_0: There exists $y \in R^n$, $y \geqslant 0$, $y_i(Ax_0 - b)_i = 0$, $i = 1, \ldots, n$, such that $c = A^* y$.

Condition a) in Theorem 11.3 is usually known as Slater's condition. There are also other nondegeneracy conditions which are less restrictive, though far more difficult to check.

The linear programming problem in the finite-dimensional case was first investigated by Kuhn and Tucker, among others, and condition (11.6) is therefore often called the Kuhn-Tucker theorem.

To conclude this lecture, we consider the extremum problem in a formulation generalizing both (11.1) and (11.4):

$$\begin{gathered} \min F_0(x), \\ F_i(x) \leqslant 0, \quad i = 1, \ldots, k, \\ F_i(x) = 0, \quad i = k+1, \ldots, n, \\ x \in Q, \end{gathered} \tag{11.8}$$

where, apart from equality and inequality constraints, there is an additional constraint which is not necessarily determined by a functional.

Theorem 11.4. Let x_0 be a solution of (11.8); assume that the functionals $F_i(x)$, $i = 0, \ldots, n$, are differentiable in a neighborhood of x_0, Q is convex and $Q^0 \neq \emptyset$. Then there exist numbers λ_i, $i = 0, \ldots, n$, not all zero, $\lambda_i \geqslant 0$, $i = 0, \ldots, k$, $\lambda_i F_i(x_0) = 0$, $i = 1, \ldots, k$, such that

$$f = \sum_{i=0}^{n} \lambda_i F_i'(x_0)$$

is a supporting functional for Q at the point x_0 (i.e., $(f, x_0) \leqslant (f, x)$ for all $x \in Q$). If $k = n$ or Q is a polyhedron, then the requirement that Q contain interior points may be relaxed.

The proof follows the same lines as before, except that here one must also use Theorem 10.5 and the result of Example 10.4.

Lecture 12

PROBLEM OF OPTIMAL CONTROL. LOCAL MAXIMUM PRINCIPLE

Problem 12.1. One of the simplest versions of the problem of optimal control is as follows: Determine functions x(t) (the phase trajectory) and u(t) (the control) satisfying the differential equation

$$\frac{dx(t)}{dt} = \varphi(x(t), u(t), t) \tag{12.1}$$

with boundary conditions

$$x(0) = c, \tag{12.2}$$

$$x(T) = d, \tag{12.3}$$

in such a way as to minimize the integral functional

$$\int_0^T \Phi(x(t), u(t), t)\,dt \tag{12.4}$$

where the control satisfies constraints of the type

$$u(t) \in M \text{ for almost all } 0 \leqslant t \leqslant T. \tag{12.5}$$

Here $x \in R^n$, $u \in R^r$, $M \subset R^r$, $\varphi(x, u, t)$ is a vector-valued function and $\Phi(x, u, t)$ a scalar function. There are other versions of the problem of optimal control, differing from the above in the form of the boundary conditions and/or the constraints imposed on the control. In some problems, T is not fixed, in others there may be constraints imposed on the phase coordinates. Later we shall consider some of these variants. It is clear that the case in which the solution is stipulated to have domain of definition (t_0, T), $t_0 \neq 0$, may be reduced to the above case by a simple substitution of variables.

We remark that the special case for which $n = r = 1$, $\varphi(x, u, t) = u$, $M = R^1$ is the classical problem of the calculus of variations:

$$\min \int_0^T \Phi(x(t), u(t), t)\, dt, \tag{12.6}$$

$$\frac{dx(t)}{dt} = u(t), \quad x(0) = c, \quad x(T) = d.$$

Before proceeding to a derivation of necessary conditions for an extremum, we formulate the problem in a more rigorous manner. As the admissible controls we take the class of bounded measurable functions (i.e., we are assuming that $u(t) \in L_\infty^{(r)}(0, T)$). Moreover, instead of equation (12.1) we shall consider the equivalent integral equation (i.e., a solution will be a pair $x(t) \in C^{(n)}(0, T)$, $u(t) \in L_\infty^{(r)}(0, T)$ satisfying the integral equation

$$x(t) = c + \int_0^t \varphi(x(\tau), u(\tau), \tau)\, d\tau).$$

Theorem 12.1 (Local maximum principle). Let $\varphi(x, u, t)$ and $\Phi(x, u, t)$ be continuous in x and u, measurable in t, continuously differentiable with respect to x, u, and moreover let

$$\varphi_x(x, u, t), \quad \varphi_u(x, u, t), \quad \Phi_x(x, u, t), \quad \Phi_u(x, u, t)$$

be bounded for all bounded x, u. Let M be a closed convex set in R^r such that $M^0 \neq \emptyset$. Let $x^0(t), u^0(t)$ be a solution of the problem (12.1).

Then there exist a number $\lambda_0 \geqslant 0$ and a function $\psi(t)$, satisfying the equation

$$\frac{d\psi(t)}{dt} = -\varphi_x{}^*(x^0(t), u^0(t), t)\, \psi(t) + \lambda_0 \Phi_x(x^0(t), u^0(t), t), \tag{12.7}$$

such that λ_0 and $\psi(t)$ cannot both be (identically) zero, and moreover

$$(-\varphi_u^*(x^0(t), u^0(t), t)\psi(t) + \lambda_0\Phi_u(x^0(t), u^0(t), t), u - u^0(t)) \geqslant 0 \tag{12.8}$$

for almost all $0 \leqslant t \leqslant T$ and all $u \in M$ (φ_x^* and φ_u^* denote the transposes of the matrices φ_x and φ_u, respectively).

Proof. Let the initial space E be the set of all pairs

$$x \in C^{(n)}(0, T), \quad u \in L_\infty^{(r)}(0, T) \ \text{(i. e.} \ E = C \times L_\infty).$$

Let Q_2 denote the set of all x, $u \in E$ satisfying equations (12.1), (12.2), (12.3), and Q_1 the set of pairs satisfying (12.5). Our problem is thus to minimize the integral

$$F(x, u) = \int_0^T \Phi(x(t), u(t), t)\, dt$$

on the set $Q = Q_1 \cap Q_2$.

We now proceed to analyze the problem in accordance with our general scheme. We shall regard Q_2 as an equality constraint and Q_1 as an inequality constraint. We first determine the corresponding cones, and then calculate the dual cones.

a) Analysis of the functional. Functionals of the type F(x, u) were considered in Examples 7.3, 7.7. The result is that $\bar{x}, \bar{u}$ lies in the cone of decrease K_0 if

$$\int_0^T [(\Phi_x(x^0, u^0, t), \bar{x}) + (\Phi_u(x^0, u^0, t), \bar{u})]\, dt < 0.$$

By Theorem 10.2, if $K_0 \neq \emptyset$, then, for any $f_0 \in K_0^*$,

$$f_0(\bar{x}, \bar{u}) = -\lambda_0 \int_0^T [(F_x(x^0, u^0, t), \bar{x}) + (F_u(x^0, u^0, t), \bar{u})]\, dt, \ \lambda_0 \geqslant 0. \tag{12.9}$$

b) Analysis of the constraint Q_1. The set of functions u(t) satisfying the constraint (12.5) is a closed convex set Q_1' in the space L_∞, and $(Q_1')^0 \neq \emptyset$ (by virtue of the assumptions of the theorem concerning the set M). Therefore the set $Q_1 = C \times Q_1'$ is also closed and convex in E, and $Q_1^0 = C \times (Q_1')^0 \neq \emptyset$. Let K_1 be the cone of feasible directions for Q_1 at the point x^0, u^0. Then, if $f_1 \in K_1^*$, it follows that $f_1 = (0, f_1')$, where $f_1' \in L_\infty^*$ is a support to Q_1' at the point u^0 (see Theorem 10.5).

c) Analysis of the constraint Q_2. This constraint was studied in Example 9.3, where the following result was obtained. Assume that the nondegeneracy condition 9.2 holds (i.e., we assume that $\varphi_u^*(x^0, u^0, t)\,\psi(t) \not\equiv 0$ for any nonzero solution $\psi(t)$ of the system

$$\frac{d\psi}{dt} = -\varphi_x^*(x^0, u^0, t)\,\psi(t)).$$

Then the tangent subspace K_2 consists of all pairs $\bar{x}, \bar{u}$ such that

$$\frac{d\bar{x}}{dt} = \varphi_x(x^0, u^0, t)\,\bar{x} + \varphi_u(x^0, u^0, t)\,\bar{u}, \quad \bar{x}(0) = 0, \tag{12.10}$$

$$\bar{x}(T) = 0. \tag{12.11}$$

Let $L_1 \subset E$ denote the set of all pairs $\bar{x}, \bar{u}$ satisfying equation (12.10), and $L_2 \subset E$ the set of pairs satisfying (12.11). Then L_1 and L_2 are subspaces and $K_2 = L_1 \cap L_2$. It is obvious that if $f \in L_2^*$, then $f(\bar{x}, \bar{u}) = (\bar{x}(T), a)$, $a \in R^n$ (since L_2 is defined by n linear functionals: $\bar{x}_i(T) = 0$). L_2^* is therefore finite-dimensional (n-dimensional), and so (Lemma 2.6) $L_1^* + L_2^*$ is closed, hence also weakly* closed (see Lecture 2). By the Corollary to Lemma 5.8, it follows that $K_2^* = L_1^* + L_2^*$. Since L_1 is a subspace, it follows from Theorem 10.1 that, for any $f_2 \in L_1^*$, we have $f_2(\bar{x}, \bar{u}) = 0$ for all $\bar{x}, \bar{u}$ satisfying (12.10). Finally, as already mentioned, if $f_3 \in L_2^*$ then $f_3(\bar{x}, \bar{u}) = (\bar{x}(T), a)$, $a \in R^n$.

d) Euler equation. Application of Theorem 6.1 to our problem implies that there exist $f_0, f_1, f_2, f_3 \in E^*$, not all zero, such that, for all $\bar{x}, \bar{u} \in E$,

$$f_0(\bar{x}, \bar{u}) + f_1(\bar{x}, \bar{u}) + f_2(\bar{x}, \bar{u}) + f_3(\bar{x}, \bar{u}) = 0, \tag{12.12}$$

where $f_0(\bar{x}, \bar{u})$ is given by (12.9), $f_1(\bar{x}, \bar{u}) = f_1'(\bar{u})$ is a support to Q_1' at u^0, $f_2(\bar{x}, \bar{u})$ vanishes for $\bar{x}, \bar{u}$ satisfying equation (12.10), and $f_3(\bar{x}, \bar{u}) = (a, \bar{x}(T))$.

e) Analysis of the Euler equation. Equation (12.12) must hold for arbitrary $\bar{x}, \bar{u}$. Let $\bar{u}$ be arbitrary, and determine a solution $\bar{x} = \bar{x}(\bar{u})$ of equation (12.10). With this choice of $\bar{x}$ and $\bar{u}$, we have $f_2(\bar{x}, \bar{u}) = 0$, and so condition (12.12) becomes

$$f_1'(\bar{u}) = \lambda_0 \int_0^T [(\Phi_x(x^0, u^0, t), \bar{x}) + (\Phi_u(x_0, u_0, t), \bar{u})]\, dt - (a, \bar{x}(T)). \tag{12.13}$$

We now transform the expression in the right-hand side of (12.13) in such a way as to replace $\bar{x}$ by $\bar{u}$. Let $\psi(t)$ be a solution of the system (12.7) with the boundary condition $\psi(T) = a$. Then, integrating by parts and using the fact that $\bar{x}, \bar{u}$ satisfy equation (12.10), we get

$$\lambda_0 \int_0^T (\Phi_x(x^0, u^0, t), \bar{x})\, dt - (a, \bar{x}(T)) =$$

$$= \int_0^T \Big(\frac{d\psi}{dt} + \varphi_x^*(x^0, u^0, t)\psi, \bar{x}\Big)\, dt - (a, \bar{x}(T)) =$$

$$= \psi\bar{x}\Big|_0^T - \int_0^T \Big(\psi, \frac{d\bar{x}}{dt}\Big)\, dt + \int_0^T (\psi, \varphi_x(x^0, u^0, t)\bar{x})\, dt - (a, \bar{x}(T)) =$$

$$= -\int_0^T (\psi, \varphi_u(x^0, u^0, t)\bar{u})\, dt = -\int_0^T (\varphi_u^*(x^0, u^0, t)\psi, \bar{u})\, dt.$$

Hence condition (12.12) becomes

$$f_1'(\bar{u}) = \int_0^T (-\varphi_u^*(x^0, u^0, t)\psi + \lambda_0\Phi_u(x^0, u^0, t), \bar{u})\, dt,$$

where $\bar{u}$ is arbitrary and $f_1^!(\bar{u})$ is a support to $Q_1^!$ at the point u^0. Now, using the result of Example 10.5 concerning the general form of a linear integral supporting functional of $Q_1^!$, we get

$$(-\varphi_u^*(x^0, u^0, t)\psi(t) + \lambda_0\Phi_u(x^0, u^0, t), u - u^0(t)) \geqslant 0$$

for almost all $0 \leqslant t \leqslant T$ and all $u \in M$, i.e., inequality (12.8) is satisfied.

Under these assumptions, the case $\lambda_0 = 0$, $\psi(t) \equiv 0$ cannot occur, since then we would have $f_0(\bar{x}, \bar{u}) \equiv 0$, $a = \psi(T) = 0$, but then $f_1^!(\bar{u}) = 0$ by (12.13), i.e., $f_0 = 0$, $f_1 = 0$, $f_3 = 0$. Equation (12.12) would then imply also $f_2 = 0$, which contradicts the assumption that there exist f_i, $i = 0, 1, 2, 3$, not all zero.

f) Analysis of exceptional cases. In the course of the proof, we made two additional assumptions: first, we assumed that $K_0 \neq \emptyset$, and, second, that the system (12.10) is nondegenerate. We shall now show that these assumptions are superfluous. If $K_0 = \emptyset$, then

$$\int_0^T [(\Phi_x(x^0, u^0, t), \bar{x}) + (\Phi_u(x^0, u^0, t), u)]\, dt = 0$$

for all $\bar{x}, \bar{u}$. Take $\lambda_0 = 1$, $\varphi(T) = 0$. Then, as before,

$$\int_0^T (\Phi_x(x^0, u^0, t), \bar{x})\, dt = -\int_0^T (\varphi_u^*(x^0, u^0, t)\psi, \bar{u})\, dt,$$

and therefore

$$\int_0^T (-\varphi_u^*(x^0, u^0, t)\psi + \Phi_u(x^0, u^0, t), \bar{u})\, dt = 0$$

for all $\bar{u}$. Hence $-\varphi_u^*(x^0, u^0, t)\psi + \Phi_u(x^0, u^0, t) = 0$ for almost all t, i.e., inequality (12.8) is satisfied. If the system (12.10) is degenerate, there exists a nonzero $\psi(t)$ which is a solution of equation (12.7) for $\lambda_0 = 0$, such that $-\varphi_u^*(x^0, u^0, t)\psi(t) \equiv 0$, which is again inequality (12.8) (with $\lambda_0 = 0$).

The proof of Theorem 12.1 is thus complete for all cases.

Remark 1. Introduce the function

$$H(x, u, \psi, t) = (\varphi(x, u, t), \psi(t)) - \lambda_0 \Phi(x, u, t).$$

Then

$$H_u(x^0, u^0, \psi, t) = \varphi_u^*(x^0, u^0, t), \psi(t)) - \lambda_0 \Phi_u(x^0, u^0, t),$$

and since a necessary condition for $H(x^0, u, \Psi, t)$ to have a maximum on M, as a func- function of u, is that $-H_u(x^0, u^0, \Psi, t)$ be a support to M at the point $u^0(t)$, it follows that (12.8) may be rephrased as follows. If x^0, u^0 is a solution of Problem 12.1 and the assumptions of Theorem 12.1 are satisfied, then $H(x^0, u, \Psi, t)$, as a function of u on M, satisfies the necessary conditions for a maximum for almost all $0 \leq t \leq T$ at the point $u = u^0(t)$. A comparison of this statement with the maximum principle (see Theorem 13.1 below) justifies the designation "local maximum principle".

Remark 2. If it is known in addition that the linearized system (12.10) is nondegenerate, then, as is easily seen by a slight modification of the proof, we can assume that $\lambda_0 = 1$.

We now discuss the modifications of the extremum conditions for different versions of the problem of optimal control. We first replace the end condition (12.3) by a more general constraint

$$G_i(x(T)) = 0, \quad i = 1, \dots, k, \tag{12.14}$$

where $G_i(x)$ are differentiable scalar functions on R^n. Then, reasoning exactly as in Example 9.3, we see that if the nondegeneracy condition holds, the tangent subspace K_2 consists of all pairs $\bar{x}, \bar{u}$ such that

$$\frac{d\bar{x}}{dt} = \varphi_x(x^0, u^0, t)\,\bar{x} + \varphi_u(x^0, u^0, t)\,\bar{u}, \quad \bar{x}(0) = 0,$$

$$(G_i{}'(x^0(T)), \bar{x}(T)) = 0, \quad i = 1, \ldots, k.$$

Now assume that $\psi(t)$ satisfies the differential equation (12.7) with the boundary condition

$$\psi(T) = \sum_{i=1}^{k} \lambda_i\, G_i{}'(x^0(T)), \tag{12.15}$$

where λ_i are arbitrary numbers. One can then use exactly the same arguments as in the proof of (12.8). Thus, in the case under consideration, the formulation of the necessary extremum conditions can be retained in the form (12.8), with the sole difference that the function $\psi(t)$ figuring therein must also satisfy the boundary conditions (12.15).

In general, if $x(t)$ is subject to boundary conditions in general form, $x(0) \in S_1$, $x(T) \in S_2$, where S_1 and S_2 are smooth manifolds, then condition (12.8) can be retained, but with additional assumptions concerning $\psi(0)$ and $\psi(T)$: $\psi(0)$ must be transversal (i.e., orthogonal to the tangent subspace) to S_1 at the point $x^0(0)$, and $\psi(T)$ transversal to S_2 at $x^0(T)$.

In particular, if the left endpoint is fixed, $x(0) = c$, while the right endpoint is free, then the boundary conditions need be imposed only on $\psi(T)$, in the form $\psi(T) = 0$. An important observation here is that in this special case the nondegeneracy condition is superfluous, and so we can assume that $\lambda_0 = 1$ (cf. Examples 9.2 and 9.3).

Analogous modifications of the extremum condition hold for the case in which S_1 and S_2 are convex sets. $\psi(0)$ must then be a support to S_1 at the point $x^0(0)$ and $\psi(T)$ a support to S_2 at $x^0(T)$.

Finally, if the functional to be minimized is not an integral functional (12.4) but a functional $\Phi_0(x(T))$, where $\Phi_0(x)$ is a function differentiable on R^n, then this problem reduces to Problem 12.1, since

$$\Phi_0(x(T)) = \Phi_0(x(0)) + \int_0^T (\Phi_0'(x(t)), \varphi(x, u, t))\, dt.$$

However, problems of this type may also be considered directly, since the cone of directions of decrease for this functional is very simply determined. It turns out that the extremum conditions have the same form; one need only take $\Phi(x, u, t) \equiv 0$ and $\psi(x(T)) = -\Phi_0'(x^0(T))$.

Some other classes of problems of optimal control (with variable T and constraints on the phase coordinates) will be considered later.

We now apply condition (12.8) to the simplest problem of optimal control — the classical variational problem (12.6). Here we have $\varphi_u(x, u, t) \equiv 1$, so that the nondegeneracy condition is satisfied and, in accordance with Remark 2, inequality (12.8) becomes

$$(-\psi(t) + \Phi_u(x^0, u^0, t))(u - u^0(t)) \geqslant 0$$

for all $u \in R^1$ and almost all $0 \leqslant t \leqslant T$. This is possible only if $-\psi(t) + \Phi_u(x^0, u^0, t) = 0$ for almost all $0 \leqslant t \leqslant T$, and hence for all $0 \leqslant t \leqslant T$, since $\psi(t)$ is continuous. Differentiating this equality with respect to t (this is legitimate, because $\psi(t)$ is differentiable), we get

$$-\frac{d\psi}{dt} + \frac{d}{dt}\Phi_u(x^0, u^0, t) = 0.$$

But it follows from (12.7) that

$$\frac{d\psi}{dt} = \Phi_x(x^0, u^0, t)$$

and hence the final result

$$-\Phi_x(x^0, u^0, t) + \frac{d}{dt}\Phi_u(x^0, u^0, t) = 0 \qquad (12.16)$$

for almost all $0 \leqslant t \leqslant T$. This is precisely the classical Euler equation, well known from the calculus of variations. Thus, inequality (12.8) is a generalization of the classical Euler equation.

If the above equality is not differentiated with respect to t, we get the extremum condition in the following form:

$$\Phi_u(x^0, u^0, t) - \int_0^t \Phi_x(x^0, u^0, \tau)\, d\tau = \psi(0) \qquad (12.17)$$

for almost all $0 \leqslant t \leqslant T$.

Hence it follows that $\Phi_u(x^0, u^0, t)$ is continuous in t, i.e.,

$$\Phi_u(x^0(t), u^0(t-0), t-0) = \Phi_u{}^0(x^0(t), u^0(t+0), t+0).$$

This is the Weierstrass-Erdmann condition from the calculus of variations.

Lecture 13

PROBLEM OF OPTIMAL CONTROL.

MAXIMUM PRINCIPLE

The result of the preceding lecture gives necessary conditions for a local minimum in the space L_∞ (with u variable). In the classical calculus of variations, such conditions are known as weak extremum conditions (they are obtained by comparing an optimal trajectory with trajectories in a weak neighborhood, i.e., trajectories which approach the optimal one uniformly in both x and u).

However, in many cases it is more natural to consider the problem in more extensive classes of spaces. For example, if the set M (see the constraint (12.5)) consists of exactly two points and $u^0(t)$ is an arbitrary function satisfying (12.5), then there are no trajectories close to $u^0(t)$ (in the L_∞-norm) which also satisfy (12.5). On the other hand, if the control $u^0(t)$ receives an increment $\bar{u}(t)$ which is nonzero only over a small time interval — known as a "spike variation" (see Fig. 11) — then the increment to the phase coordinates and the functional will also be small (even when $\|\bar{u}\|_{L_\infty}$ is not small). It is therefore desirable to consider the problem in a space for which the norm of the spike variation is small (for example, in L_1). The usual term for conditions for a local extremum in such spaces is strong extremum conditions.

We shall now consider the derivation of conditions of this type for optimal control problems. In so doing we shall be able to relax the condition that M be convex and contain interior points; neither shall we need the assumption that $\Phi(x, u, t)$ and $\varphi(x, u, t)$ are differentiable with respect to u. Instead of directly analyzing the functional and constraints in the space L_1, we shall employ an artificial method, first suggested by Dubovitskii and Milyutin. The main idea of the method is to introduce a variable time transformation; this will be done in such a way that a small variation of the transformation corresponds to a spike variation in the initial problem.

We proceed to the rigorous formulation. We first formulate the problem to be solved.

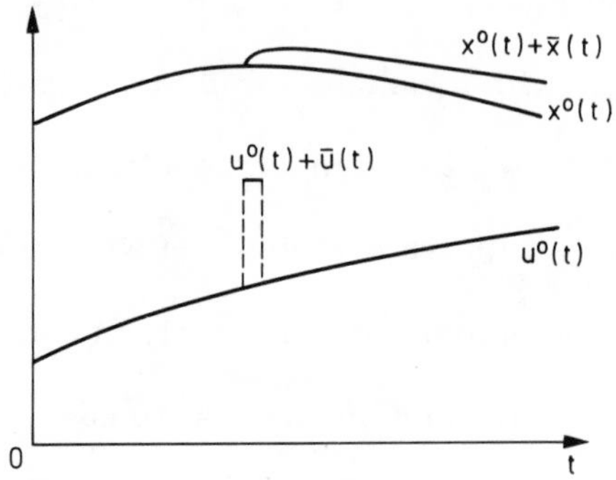

Fig. 11.

Problem 13.1. Minimize

$$F(x, u) = \int_{t_0}^{t_1} \Phi(x(t), u(t))\, dt \tag{13.1}$$

under the constraints

$$\frac{dx(t)}{dt} = \varphi(x(t), u(t)), \quad x(t_0) = c \tag{13.2}$$

$$x(t_1) = d, \tag{13.3}$$

$$u(t) \in M \subset R^r \text{ for almost all } t_0 \leqslant t \leqslant t_1. \tag{13.4}$$

The functional is to be minimized over all $t_1 \geqslant t_0$, all bounded measurable u(t) and all continuous x(t) satisfying (13.2) almost everywhere. We shall assume that the following condition holds.

Condition 13.1. The functions $\Phi(x, u)$ and $\varphi(x, u)$ are defined on R^n, with values in R^1 and R^n, respectively; they are continuous in u and continuously differentiable in x; $\Phi_x(x, u)$ and $\varphi_x(x, u)$ are bounded for arbitrary bounded x, u.

Note the differences between Problem 13.1 with Condition 13.1 and the problem (12.1) considered in the preceding lecture. First, we assumed there that M was convex and $M^0 \neq \emptyset$, whereas here M is any set in R^r. Second, Condition 13.1 does not require that $\Phi(x, u)$ and $\varphi(x, u)$ be differentiable with respect to u; this is natural, in view of the fact that M may consist of finitely many points. Third, the functions $\Phi(x, u)$ and $\varphi(x, u)$ in Problem 13.1 do not depend explicitly on t (so that we have what is known as a <u>time-invariant system</u>). The case of time-varying systems will be examined somewhat later. Fourth, the time t_1 in Problem 13.1 is not fixed, in contradistinction to Problem 12.1.

We introduce a time transformation $t \longrightarrow \tau$, mapping $[t_0, t_1]$ onto $[0, 1]$, defined by a certain function $v(\tau)$:

$$t(\tau) = t_0 + \int_0^\tau v(\tau)\, d\tau, \tag{13.5}$$

$$t(1) = t_1, \tag{13.6}$$

$$v(\tau) \geqslant 0. \tag{13.7}$$

The transformation thus defined is one-to-one if $v(\tau) > 0$, $0 \leqslant \tau \leqslant 1$. But if $v(\tau)$ vanishes over some closed interval Δ, then $t(\tau)$ = const for $\tau \in \Delta$, so that one value of t will correspond to an entire interval on the τ-axis (Fig. 12). To make the definition of $\tau(t)$ one-to-one, we shall assume that

$$\tau(t) = \inf\{\tau\colon t(\tau) = t\}. \tag{13.8}$$

We now formulate a new problem.

P r o b l e m 13.2. Minimize the functional

$$F(x, u, v) = \int_0^1 v(\tau)\, \Phi(x(\tau), u(\tau))\, d\tau \tag{13.9}$$

for

$$x(\tau) \in C^{(n)}(0, 1), \quad u(\tau) \in L_\infty^{(r)}(0, 1), \quad v(\tau) \in L_\infty^{(1)}(0, 1)$$

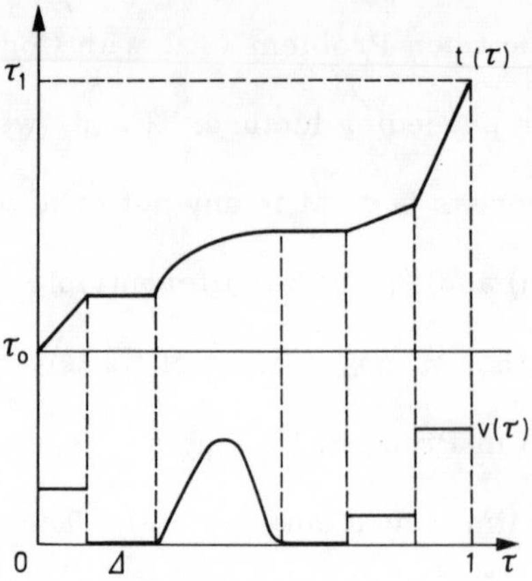

Fig. 12.

under the constraints

$$\frac{dx(\tau)}{d\tau} = v(\tau)\,\varphi(x(\tau), u(\tau)), \tag{13.10}$$

$$x(1) = d, \tag{13.11}$$

$$v(\tau) \geqslant 0, \tag{13.12}$$

$$u(\tau) \in M \text{ for almost all } 0 \leqslant \tau \leqslant 1. \tag{13.13}$$

Let us determine the relation between Problems 13.1 and 13.2.

L e m m a 13.1. Let x(t) and u(t) satisfy equation (13.2) and the constraints (13.3), (13.4). Then, for any function $v(\tau)$ satisfying (13.6) and (13.7), the functions $x(\tau) = x(t(\tau))$ and

$$u(\tau) = \begin{cases} u(t(\tau)) \text{ for } \tau \in R_1, \\ \text{arbitrary for } \tau \in R_2 \end{cases} \tag{13.14}$$

(where $t(\tau)$ is defined by (13.5),

$$R_1 = \{\tau \in [0,1]\colon v(\tau) > 0\}, \quad R_2 = \{\tau \in [0,1]\colon v(\tau) = 0\})$$

satisfy equation (13.10) and the constraints (13.11)—(13.13); moreover, F(x, u) = F(x, u, v).

Conversely, if $x(\tau)$, $u(\tau)$, $v(\tau)$ satisfy (13.10)—(13.13), then the functions defined by $x(t) = x(\tau(t))$, $u(t) = u(\tau(t))$, with $t_1 = t_0 + \int_0^1 v(\tau)\,d\tau$ (where $\tau(t)$ is given by (13.8)) satisfy (13.2)—(13.4), and moreover $F(x, u) = F(x, u, v)$.

Corollary. If $x^0(t)$, $u^0(t)$ is a solution of Problem 13.1, then, for any function $v^0(\tau)$ satisfying (13.6)—(13.7), the functions $x^0(\tau)$, $u^0(\tau)$, $v^0(\tau)$ defined by (13.14) furnish a solution of Problem 13.2. Of course, this also furnishes a solution of a new Problem 13.3, which is the same as Problem 13.2 except that $u^0(\tau)$ is fixed and $F(x, u^0, v)$ is to be minimized only over $x(\tau)$ and $v(\tau)$.

The proof of the lemma is almost self-evident. Wherever $v(\tau) > 0$, both sides of equation (13.10) may be divided by $v(\tau)$; one then uses the fact that

$$\frac{dx(t)}{dt} = \frac{dx(t(\tau))}{dt}\Big/\frac{d(t(\tau))}{d\tau} = \frac{dx(\tau)}{v(\tau)\,d\tau},$$

so that (13.10) is transformed into (13.2). But if $v(\tau) = 0$, then

$$\frac{dx(\tau)}{d\tau} = 0, \quad x(\tau) = \text{const} = x(\tau(t)).$$

Analogous arguments will show that the functionals $F(x, u)$ and $F(x, u, v)$ are identical.

Thus, let $x^0(t)$, $u^0(t)$, t_1 be a solution of the initial problem. Let $v^0(\tau)$ be some function satisfying (13.6) and (13.7) — the form of the function will be specified later. Define $u^0(\tau)$ in accordance with (13.14) — the choice of $u^0(\tau)$ for $\tau \in R_2$ will also be specified later. Let us solve Problem 13.3, i.e.,

$$\min_{x,v} F(x, u^0, v) = \int_0^1 v(\tau)\,\Phi(x(\tau), u^0(\tau))\,d\tau, \tag{13.15}$$

$$\frac{dx(\tau)}{d\tau} = v(\tau)\,\varphi(x(\tau), u^0(\tau)), \tag{13.16}$$

$$x(1) = d, \tag{13.17}$$

$$v(\tau) \geqslant 0. \tag{13.18}$$

By the Corollary to Lemma 13.1, $x^0(\tau)$ and $v^0(\tau)$ constitute a solution. To derive necessary conditions for a minimum in Problem 13.3, we can use the results of the preceding lecture, with $v(\tau)$ playing the role of control. These results are indeed applicable to the problem, since, first, the constraint (13.18) is of the form $v(\tau) \in M_1$, $0 \leqslant \tau \leqslant 1$, where M_1 is a convex set in R^1 with an interior point (the positive half-axis), and, second, the integrand and the right-hand side of the equation are differentiable with respect to v, and, third, the time T = 1 is fixed.

Therefore, using the local maximum principle (Theorem 12.1) for Problem 13.3, we see that there exist $\psi(\tau)$, $\lambda_0 \geqslant 0$, not both zero, such that

$$\frac{d\psi(\tau)}{d\tau} = -v^0(\tau)\varphi_x^*(x^0(\tau), u^0(\tau))\psi(\tau) + \lambda_0 v^0(\tau)\Phi(x^0(\tau), u^0(\tau)), \tag{13.19}$$

$$[-(\varphi(x^0(\tau), u^0(\tau)), \psi(\tau)) + \lambda_0\Phi(x^0(\tau), u^0(\tau))](v - v^0(\tau)) \geqslant 0 \tag{13.20}$$

for almost all $0 \leqslant \tau \leqslant 1$ and all $v \geqslant 0$. It follows from the last inequality that

$$-(\varphi(x^0(\tau), u^0(\tau)), \psi(\tau)) + \lambda_0\Phi(x^0(\tau), u^0(\tau)) = 0 \text{ for almost all } \tau \in R_1, \tag{13.21}$$

$$-(\varphi(x^0(\tau)), u^0(\tau)), \psi(\tau)) + \lambda_0\Phi(x^0(\tau), u^0(\tau)) \geqslant 0 \text{ for almost all } \tau \in R_2 \tag{13.22}$$

$$(\text{here } R_1 = \{\tau\colon v^0(\tau) > 0\},\quad R_2 = \{\tau\colon v^0(\tau) = 0\}).$$

We now introduce the function $\psi(t) = \psi(\tau(t))$. Since

$$\psi(t) = \text{const} = \psi(\tau(t))$$

wherever $v^0(\tau) = 0$, it follows that $\psi(t)$ satisfies the differential equation

$$\frac{d\psi(t)}{dt} = -\varphi_x^*(x^0(t), u^0(t))\psi(t) + \lambda_0\Phi(x^0(t), u^0(t)), \quad t_0 < t < t_1. \tag{13.23}$$

If we now replace τ by t in condition (13.21), we get

$$-(\varphi(x^0(t), u^0(t)), \psi(t)) + \lambda_0 \Phi(x^0(t), u^0(t)) = 0 \qquad (13.24)$$

for almost all $t_0 \leq t \leq t_1$ such that $t = t(\tau)$, $v^0(\tau) > 0$. But the set of points t such that $t = t(\tau)$, $\tau \in R_1$, is at most countable (for the monotone function

$$t(\tau) = t_0 + \int_0^\tau v^0(\tau)\, d\tau$$

is one-to-one everywhere except, perhaps, on a countable set of t; see Natanson [1]). Therefore condition (13.24) holds for almost all $t_0 \leq t \leq t_1$.

Before proceeding to an analysis of (13.22), recall that we have still to specify the form of the function $v^0(\tau)$ and the values of $u^0(\tau)$ on the set R_2. This we now do. Let R_1 be a perfect nowhere dense subset of [0, 1] (i.e., a set, containing no intervals, obtained from [0, 1] by deleting a countable number of intervals). Assume moreover that the intersection of R_1 with any interval, if not empty, has positive measure. The construction of such a set is analogous to that of the Cantor set, except that the intervals must be deleted in such a way that the measure of the residual set is always greater than some positive constant. We now define $v^0(\tau)$:

$$v^0(\tau) = \begin{cases} \dfrac{t_1 - t_0}{\mu(R_1)}, & \tau \in R_1, \\ 0, & \tau \in R_2 = [0,1] - R_1. \end{cases}$$

Then (13.6) and (13.7) will hold. We now define $u^0(\tau)$ on R_2. The set R_2, as the complement of the perfect set R_1, is the union of a countable set of intervals. Let $\tau \in \Delta$, where Δ is one of these intervals. Divide Δ into a countable number of disjoint intervals, closed on the left: $\Delta = \bigcup_{i=1}^{\infty} \Delta_i$. Take elements $u^i \in M$, $i = 1, 2, \ldots, \infty$ which constitute a countable dense subset of M (this is always possible, for R^r is separable). Set $u^0(\tau) = u^i$ for $\tau \in \Delta_i$. This completes the definition of $v^0(\tau)$ and $u^0(\tau)$.

We can now go on to our analysis of condition (13.22). Writing this inequality for $\tau \in \Delta_i \subset \Delta \subset R_2$ (with the same notation as before), we get

$$-(\varphi(x^0(\tau), u^i), \psi(\tau)) + \lambda_0 \Phi(x^0(\tau), u^i) \geqslant 0.$$

But

$$x^0(\tau) = \text{const} = x^0(\tau(t)), \quad \psi(\tau) = \text{const} = \psi(\tau(t))$$

for all $\tau \in \Delta$. Thus

$$-(\varphi(x^0(\tau(t)), u^i), \psi(\tau(t))) + \lambda_0 \Phi(x^0(\tau(t)), u^i) \geqslant 0, \quad i = 1, 2, \ldots, \infty.$$

Since $\{u^i\}$ is dense in M, it follows that

$$-(\varphi(x^0(\tau(t)), u), \psi(\tau(t))) + \lambda_0 \Phi(x^0(\tau(t)), u) \geqslant 0, \tag{13.24*}$$

for all $u \in M$ and almost all $\tau \in R_2$.

Setting $x^0(t) = x^0(\tau(t))$, $\Psi(t) = \Psi(\tau(t))$, we get

$$-(\varphi(x^0(t), u), \psi(t)) + \lambda_0 \Phi(x^0(t), u) \geqslant 0 \quad \text{for all } u \in M, \tag{13.25}$$

and this inequality is valid for all t such that the measure of the set $R_t = \{\tau \in R_2 : t(\tau) = t\}$ is positive. Indeed, if $\mu(R_t) > 0$, then inequality (13.24*) holds for some $\tau \in R_t$ (since (13.24*) can fail to hold only for a set of measure zero). Now the set consisting of all these points t is dense in $[t_0, t_1]$. In fact, if it contains some segment, then its complete preimage under the mapping $t(\tau)$ is also a segment. But the set R_2 is dense, and its intersection with this segment is therefore not empty and contains a segment Δ (for R_2 is the union of a countable set of intervals). Let $\tau \in \Delta$, $t = t(\tau)$; then $R_t \supset \Delta$, so that $\mu(R_t) > 0$ and we have a contradiction. Thus the set of all t for which (13.25) holds is dense in $[t_0, t_1]$. But the left-hand side of (13.25) is a continuous function of t, and hence (13.25) is valid for all $t \in [t_0, t_1]$.

We have thus proved the following result.

Theorem 13.1 (Pontryagin maximum principle). Assume that Condition 13.1 holds for Problem 13.1, and let $x^0(t)$, $u^0(t)$, t_1 be a solution. Then there exist $\psi(t)$ and $\lambda_0 \geqslant 0$, not both zero, such that

$$\frac{d\psi}{dt} = -\varphi_x^*(x^0, u^0)\psi + \lambda_0 \Phi_x(x^0, u^0)$$

$$H(x^0(t), u^0(t), \psi(t), t) = 0 \quad \text{for almost all } t_0 \leqslant t \leqslant t_1,$$

$$H(x^0(t), u, \psi(t), t) \leqslant 0 \quad \text{for all } u \in M \text{ and almost all } t_0 \leqslant t \leqslant t_1,$$

$$\text{where } H(x, u, \psi, t) = (\varphi(x, u), \psi) - \lambda_0 \Phi(x, u).$$

In other words, the function $H(x^0(t), u, \psi(t), t)$ assumes a maximum on $u \in M$ for $u = u^0(t)$, for almost all $t_0 \leqslant t \leqslant t_1$.

As for the other variants of the optimal control problem, it can be shown that modifications in the boundary conditions on $x(t_0)$, $x(t_1)$ induce the same modifications in the conditions on $\psi(t_0)$, $\psi(t_1)$ as in the local maximum principle, and we shall therefore not repeat the relevant arguments from Lecture 12. We shall devote a little more attention to the case of time-varying problems.

Thus, we now assume that the functions Φ and φ in Problem 13.1 depend on t: $\Phi = \Phi(x, u, t)$, $\varphi = \varphi(x, u, t)$. We shall assume that $\Phi(x, u, t)$, and $\varphi(x, u, t)$ satisfy Condition 13.1 and are continuously differentiable with respect to t. This problem will be reduced to Problem 13.1. We supplement the variables $x(t) = (x_1(t), \ldots, x_n(t))$ with a new variable $x_{n+1}(t)$, satisfying the differential equation

$$\frac{dx_{n+1}(t)}{dt} = 1, \quad x_{n+1}(t_0) = t_0.$$

Then the problem becomes

$$\min \int_{t_0}^{t_1} \Phi(x(t), u(t), x_{n+1}(t))\, dt,$$

$$\frac{dx(t)}{dt} = \varphi(x(t), u(t), x_{n+1}(t)), \quad x(t_0) = c,$$

$$\frac{dx_{n+1}(t)}{dt} = 1, \; x_{n+1}(t_0) = t_0,$$

$$x(t_1) = d,$$
$$u(t) \in M.$$

Apply the maximum principle to this problem. The result is as follows. Let $x^0(t)$, $u^0(t)$, $x^0_{n+1}(t)$ be a solution. To the functions $\psi(t) = (\psi_1(t), \ldots, \psi_n(t))$ we add a new function $\psi_{n+1}(t)$, and consider the equations

$$\frac{d\psi(t)}{dt} = -\varphi_x^*(x^0(t), u^0(t), x^0_{n+1}(t))\,\psi(t) + \lambda_0 \Phi_x(x^0(t), u^0(t), x^0_{n+1}(t)),$$

$$\frac{d\psi_{n+1}(t)}{dt} = -(\varphi_t(x^0(t), u^0(t), x^0_{n+1}(t)), \psi(t)) +$$
$$+ \lambda_0 \Phi_t(x^0(t), u^0(t), x^0_{n+1}(t)), \quad \psi_{n+1}(t_1) = 0$$

(the boundary condition for $\psi_{n+1}(t_1)$ is given, since $x_{n+1}(t_1)$ is free; see above). The maximum principle then implies

$$\max_{u \in M} [H(x^0(t), u(t), \psi(t), t) + \psi_{n+1}(t)] = H(x^0(t), u^0(t), \psi(t), t) + \psi_{n+1}(t) = 0,$$

where, as before, $H(x, u, \psi, t) = (\varphi(x, u, t), \psi) - \lambda_0 \Phi(x, u, t)$. Hence, using the fact that the right-hand side of the equation for $\psi_{n+1}(t)$ is independent of ψ_{n+1}, so that the equation can be integrated, we get

$$\max_{u \in M} H(x^0(t), u, \psi(t), t) = H(x^0(t), u^0(t), \psi(t), t) =$$
$$= \int_t^{t_1} [-(\varphi_t(x^0; u^0, t), \psi(t)) + \lambda_0 \Phi_t(x^0, u^0, t)]\, dt.$$

Thus, the extremum condition for our problem has the same form as that for the time-invariant problem, except that the condition $H(x^0(t), u^0(t), \psi(t), t) = 0$ is replaced by

$$H(x^0(t),\, u^0(t),\, \psi(t),\, t) = \int_t^{t_1} [-(\varphi_t(x^0,\, u^0,\, t),\, \psi(t)) + \lambda_0 \Phi_t(x^0,\, u^0,\, t)]\, dt. \qquad (13.25^*)$$

We now consider a time-varying problem in which t_1 is fixed. The variable $x_{n+1}(t)$, introduced as before, will now satisfy the boundary condition $x_{n+1}(t_1) = t_1$. Therefore, in the adjoint equation, there will be no boundary condition on $\psi_{n+1}(t_1)$. Hence the maximum principle will have the same form, but condition (13.25) will be replaced by

$$H(x^0(t),\, u^0(t),\, \psi(t),\, t) = \lambda + \int_t^{t_1} [-(\varphi_t(x^0,\, u^0,\, t),\, \psi(t)) + \lambda_0 \Phi_t(x^0,\, u^0,\, t)]\, dt, \qquad (13.26)$$

where λ is a constant. In particular, in the case of a time-invariant problem we get the condition $H(x^0(t),\, u^0(t),\, \psi(t),\, t) \equiv \lambda$ (in contrast to the condition $H(x^0(t),\, u^0(t),\, \psi(t),\, t) = 0$ for the problem with t_1 free).

To conclude this lecture, we apply the maximum principle to the <u>classical problem of the calculus of variations</u> (12.6). Since in this case the variational equation is nondegenerate, we can take $\lambda_0 = 1$.

Therefore,

$$H(x^0,\, u,\, \psi,\, t) = u\psi - \Phi(x^0,\, u,\, t)$$

where $\psi(t)$ satisfies the differential equation

$$\frac{d\psi}{dt} = \Phi_x(x^0,\, u^0,\, t).$$

Hence

$$\psi(t) = \psi(0) + \int_0^t \Phi_x(x^0,\, u^0,\, \tau)\, d\tau = \Phi_u(x^0,\, u^0,\, t)$$

(this follows from the Euler equation (12.17)). Thus the maximum principle yields the following inequality for all u and t:

$$u^0 \psi - \Phi(x^0, u^0, t) \geqslant u \psi - \Phi(x^0, u, t).$$

Alternatively, setting

$$E(x^0, u^0, u, t) = \Phi(x^0, u, t) - \Phi(x^0, u^0, t) + (u^0 - u)\,\Phi_u(x^0, u^0, t)$$

(the Weierstrass function), we have

$$E(x^0, u^0, u, t) \geqslant 0 \tag{13.27}$$

for all u, t. This is the well-known necessary condition for a strong extremum (Weierstrass condition) from the calculus of variations. The maximum principle is thus a natural generalization of the Weierstrass condition to the problem of optimal control.

Lecture 14

PROBLEM OF OPTIMAL CONTROL. CONSTRAINTS ON PHASE COORDINATES, MINIMAX PROBLEM

In this lecture we shall consider a more general problem of optimal control, in which there are also constraints on the phase coordinates, of the type $G(x(t), t) \leqslant 0$ for all t.

We shall then carry over our results to a problem of optimal control in which the performance index is not an integral functional but a functional of the type

$$\max_t G(x(t), t).$$

Problem 14.1. Minimize

$$\int_0^T \Phi(x(t), u(t), t)\, dt \tag{14.1}$$

under the constraints

$$\frac{dx(t)}{dt} = \varphi(x(t), u(t), t), \quad x(0) = c, \tag{14.2}$$

$$x(T) = d \tag{14.3}$$

$$u(t) \in M \text{ for almost all } 0 \leqslant t \leqslant T, \tag{14.4}$$

$$G(x(t), t) \leqslant 0 \text{ for all } 0 \leqslant t \leqslant T \tag{14.5}$$

in the class of functions

$$x(t) \in C^{(n)}(0, T), \quad u(t) \in L_\infty^{(r)}(0, T).$$

The time T is fixed.

The only difference between Problem 14.1 and Problem 12.1 is the additional constraint (14.5). $G(x, t)$ is a scalar function; the other notation has the same meaning as in Problem 12.1.

Theorem 14.1. (Local maximum principle). Under the assumptions of Theorem 12.1, let $G(x, t)$ be continuous in (x, t), continuously differentiable with respect to x, $G_x(x, t) \neq 0$ when $G(x, t) = 0$ and $G(c, 0) < 0$, $G(d, T) < 0$. Let $x^0(t)$, $u^0(t)$ be a solution of Problem 14.1. Then there exist a function $\psi(t)$, a number $\lambda_0 \geq 0$, a vector $a \in R^r$ and a nonnegative measure $d\mu(t)$ whose support is the set

$$R = \{t \in [0, T]: \; G(x^0(t), t) = 0\}$$

such that $\psi(t)$ and λ_0 are not both zero, and

$$-\psi(t) = -a + \int_t^T (-\varphi_x^*(x^0, u^0, \tau)\psi(\tau) + \lambda_0 \Phi_x(x^0, u^0, \tau))\, d\tau + \int_t^T G_x(x^0(\tau), \tau)\, d\mu(\tau), \tag{14.6}$$

$$(-\varphi_u^*(x^0, u^0, t)\psi(t) + \lambda_0 \Phi_u(x^0, u^0, t), \quad u - u^0(t)) \geq 0 \tag{14.7}$$

for all $u \in M$ and almost all $0 \leq t \leq T$.

Proof. The problem may be presented as follows:

$$\min F(x, u) \text{ for } (x, u) \in Q_1 \cap Q_2 \cap Q_3 \subset E = C \times L_\infty,$$

where $F(x, u)$ is given by (14.1), Q_1 by (14.4), Q_2 by (14.5), and Q_3 by (14.2) and (14.3). $F(x, u)$, Q_1 and Q_3 have already been investigated (see the proof of Theorem 12.1) and the corresponding cones determined. We now consider the constraint Q_2. By Example 7.5 and the Corollary to Theorem 8.1, the cone of admissible directions K_2 for Q_2 consists of all $\bar{x}, \bar{u}$ such that $(G_x(x^0(t), t), \bar{x}(t)) < 0$ for all $t \in R$.

Under these conditions, condition a) of the Corollary to Theorem 8.1 holds, since $G_x(x^0(t), t) \neq 0$ on R.

A cone of the required type was considered in Example 10.3, and the dual cone was also determined there.

Using this result, we see that if $f_2 = (f_2', 0) \in K_2^*$, $f_2' \in C^*$, then there exists a measure $d\mu(t)$, nonnegative and with support on R, such that

$$f_2(\bar{x}, \bar{u}) = f_2'(\bar{x}) = -\int_R (G_x(x^0(t), t), \bar{x}(t))\, d\mu(t).$$

Continuing now exactly as in the proof of Theorem 12.1 and applying the Dubovitskii-Milyutin theorem, we get necessary conditions for an extremum:

$$-\lambda_0 \int_0^T [(\Phi_x(x^0, u^0, t), \bar{x}(t)) + (\Phi_u(x^0, u^0, t), \bar{u}(t))]\, dt + f_1'(\bar{u}) -$$
$$- \int_0^T (G_x(x^0(t), t), \bar{x}(t))\, d\mu(t) + f_3(\bar{x}, \bar{u}) + (a, \bar{x}(T)) = 0 \tag{14.8}$$

for all $\bar{x}, \bar{u}$. Here $f_1'(\bar{u})$ is a supporting functional for (14.4) at u^0, $a \in R^n$ is a vector, and $f_3(\bar{x}, \bar{u})$ vanishes for $\bar{x}, \bar{u}$ that satisfy the equation

$$\frac{d\bar{x}(t)}{dt} = \varphi_x(x^0, u^0, t)\bar{x}(t) + \varphi_u(x^0, u^0, t)\bar{u}(t),\ \bar{x}(0) = 0. \tag{14.9}$$

We now take an arbitrary $\bar{u} \in L_\infty^{(r)}(0, T)$ and find $\bar{x} = \bar{x}(\bar{u})$ as a solution of equation (14.9). Since $f_3(\bar{x}, \bar{u}) = 0$ for these $\bar{x}, \bar{u}$, it follows that (14.8) can be rewritten as

$$-\lambda_0 \int_0^T [(\Phi_x(x^0, u^0, t), \bar{x}(t)) + (\Phi_u(x^0, u^0, t), \bar{u}(t))]\, dt + f_1'(\bar{u}) -$$
$$- \int_0^T (G_x(x^0(t), t), \bar{x}(t))\, d\mu(t) + (a, \bar{x}(T)) = 0. \tag{14.10}$$

We now transform the terms in (14.10) involving $\bar{x}$, in such a way as to obtain an expression depending explicitly only on $\bar{u}$. Let $\Psi(t)$ be a solution of the integral equation (14.6). Note that, since $G(x^0(T), T) = G(d, T) < 0$, it follows that $T \notin R$; hence the measure μ does not have support at T and it follows from (14.6) that $\Psi(T) = a$. Multiply both sides of (14.6) by $\frac{d\bar{x}(t)}{dt}$ (scalar product!) and integrate from 0 to T:

$$-\int_0^T\left(\frac{d\bar{x}(t)}{dt},\ \psi(t)\right)dt=-\int_0^T\left(a,\ \frac{d\bar{x}(t)}{dt}\right)dt+$$

$$+\int_0^T\left(\frac{d\bar{x}(t)}{dt},\ \int_t^T(-\varphi_x^*(x^0,u^0,\tau)\,\psi(\tau)+\lambda_0\Phi_x(x^0,u^0,\tau))\,d\tau\right)dt+$$

$$+\int_0^T\left(\frac{d\bar{x}(t)}{dt},\ \int_t^T G_x(x^0(\tau),\tau)\,d\mu(\tau)\right)dt. \qquad (14.11)$$

Since $\bar{x}(0)=0$,

$$-\int_0^T\left(a,\ \frac{d\bar{x}(t)}{dt}\right)dt=-(a,\ \bar{x}(T)\ ).$$

Integrating the second term by parts and using equation (14.9), we get

$$\int_0^T\left(\frac{d\bar{x}(t)}{dt},\ \int_t^T(-\varphi_x^*(x^0,u^0,\tau)\,\psi(\tau)+\lambda_0\Phi_x(x^0,u^0,\tau))\,d\tau\right)dt=$$

$$=\int_0^T(\bar{x}(t),\ -\varphi_x^*(x^0,u^0,t)\,\psi(t)+\lambda_0\Phi_x(x^0,u^0,t))\,dt=$$

$$=\int_0^T(-\varphi_x(x^0,u^0,t)\,\bar{x}(t),\ \psi(t))\,dt+\lambda_0\int_0^T(\bar{x}(t),\ \Phi_x(x^0,u^0,t))\,dt=$$

$$=\int_0^T\left(-\frac{d\bar{x}(t)}{dt}+\varphi_u(x^0,u^0,t)\,\bar{u}(t),\ \psi(t)\right)dt+$$

$$+\lambda_0\int_0^T(\bar{x}(t),\ \Phi_x(x^0,u^0,t))\,dt=-\int_0^T\left(\frac{d\bar{x}(t)}{dt},\ \psi(t)\right)dt+$$

$$+\int_0^T(\varphi_u^*(x^0,u^0,t)\,\psi(t),\ \bar{u}(t))\,dt+\ \lambda_0\int_0^T(\bar{x}(t),\ \Phi_x(x^0,u^0,t))\,td.$$

Finally, transforming the last term of (14.11), using the rule for integration by parts of Stieltjes integrals (see, e.g., Natanson [1]), and also the fact that

$$G(x^0(0),0)=G(c,0)<0,\quad G(x^0(T),T)=G(d,T)<0,$$

so that $0\notin R$, $T\notin R$ and consequently the measure μ does not have support at the points 0 and T, we get

$$\int_0^T \left(\frac{d\bar{x}(t)}{dt}, \int_t^T G_x(x^0(\tau), \tau)\, d\mu\, \tau)\right) dt = \int_0^T (G_x(x^0(t), t), \bar{x}(t))\, d\mu\, (t).$$

Substituting all these expressions into (14.11) and canceling like terms, we get

$$-\lambda_0 \int_0^T (\Phi_x(x^0, u^0, t), \bar{x}(t))\, dt - \int_0^T (G_x(x^0(t), t), \bar{x}(t)) d\mu\,(t) + (a, \bar{x}(T)) =$$

$$= \int_0^T (\varphi_u{}^*(x^0, u^0, t)\, \psi(t), \bar{u}(t))\, dt.$$

Thus condition (14.10) becomes

$$\int_0^T (\lambda_0 \Phi_u(x^0, u^0, t) - \varphi_u{}^*(x^0, u^0, t)\, \psi(t), \bar{u}(t))\, dt = f_1'(\bar{u}).$$

The conclusion of the proof is exactly the same as for Theorem 12.1.

Remark 1. It is clear from equation (14.6) that the function $\Psi(t)$ is in general not continuous — it is discontinuous at every point where $\mu(t)$ has a discontinuity (i.e., at every point of support of the measure). Nevertheless, if we perform formal integration of equation (14.6) with respect to t, we get a differential equation

$$\frac{d\psi(t)}{dt} = -\varphi_x{}^*(x^0, u^0, t)\, \psi(t) + G_x(x^0, t) \frac{d\mu(t)}{dt}, \tag{14.12}$$

which differs from the corresponding equation for $\Psi(t)$ in the problem without constraints on the phase coordinates only by the presence of the additional term $G_x(x^0, t) \frac{d\mu(t)}{dt}$.

Remark 2. Just as in the problem without constraints on the phase coordinates, modification of the boundary conditions for x(t) leads to modifications in the boundary conditions for $\Psi(t)$. For example, if x(T) is free, then $\Psi(T) = 0$ (i.e., a = 0 in equation (14.6)).

Remark 3. The requirement that G(x, t) be differentiable with respect to x can be relaxed. It is sufficient that G(x, t) be differentiable with respect to x in any direction and that the directions of decrease generate an open cone (see

Example 7.6). These conditions hold for convex continuous functions (for example, $G(x, t) = |x|$).

Remark 4. If there are several constraints on the phase coordinates, $G_i(x(t), t) \leqslant 0$, $i = 1, \ldots, k$, they may be reduced to a single constraint by setting

$$G(x, t) = \max_{1 \leqslant i \leqslant k} G_i(x, t)$$

and the preceding remark then applies. However, they may also be retained as k separate inequality constraints and the Dubovitskii-Milyutin theorem applied directly.

We now present a theorem which generalizes Theorem 13.1 (the maximum principle) to the case of problems with constraints on the phase coordinates. We first formulate the problem.

Problem 14.2. Minimize the functional

$$\int_{t_0}^{t_1} \Phi(x(t), u(t))\, dt$$

with respect to $x(t) \in C^{(n)}(0, T)$, bounded measurable $u(t)$, and t_1, under the constraints

$$\frac{dx(t)}{dt} = \varphi(x(t), u(t)), \quad x(t_0) = c,$$

$$x(t_1) = d,$$

$$u(t) \in M \text{ for almost all } t_0 \leqslant t \leqslant t_1,$$

$$G(x)t)) \leqslant 0 \text{ for all } t_0 \leqslant t \leqslant t_1.$$

We use the same notation as in Problem 13.1. In general, the only difference between Problem 14.2 and Problem 13.1 is the presence of the constraint $G(x(t)) \leqslant 0$.

Theorem 14.2. (Maximum principle). Under the assumptions of Theorem

13.1, let $G(x)$ be continuously differentiable, $G_x(x) \neq 0$ when $G(x) = 0$ and $G(c) < 0$, $G(d) < 0$. Let $x^0(t)$, $u^0(t)$, t_1 be a solution of Problem 14.2. Then there exist a function $\Psi(t)$, a number $\lambda_0 \geqslant 0$, a vector $a \in R^n$ and a nonnegative measure $d\mu(t)$ with support on the set

$$R = \{t \in [t_0, t_1]: G(x^0(t)) = 0\}$$

such that

$$-\psi(t) = -a + \int_t^{t_1} [-\varphi_x^*(x^0, u^0, \tau)\psi(\tau) + \lambda_0 \Phi_x(x^0, u^0, \tau)]d\tau + \\ + \int_t^{t_1} G_x(x^0(\tau))\, d\mu(\tau),$$

and moreover the function

$$H(x, \psi, u, t) = (\varphi(x, u, t), \psi(t)) - \lambda_0 \Phi(x, u, t)$$

satisfies the following conditions:

$$H(x^0(t), \psi(t), u^0(t), t) = 0 \text{ for almost all } t_0 \leqslant t \leqslant t_1,$$
$$H(x^0(t), \psi(t), u(t), t) \leqslant 0 \text{ for almost all } t_0 \leqslant t \leqslant t_1$$

and all $u \in M$. Moreover, λ_0 and $\Psi(t)$ cannot both be (identically) zero.

We shall not present the proof of this theorem, since it follows the same lines as that of Theorem 13.1. In brief, one introduces a transformation

$$t(\tau) = t_0 + \int_{t_0}^{\tau} v(\tau)\, d\tau,$$

constructs an auxiliary problem with control $v(\tau)$, linear in $v(\tau)$, and then applies the local maximum principle (Theorem 14.1) to this problem. One then goes back from τ to t by a suitable choice of $v^0(\tau)$ and $u^0(\tau)$.

For time-varying problems (including those with constraints of the type $G(x(t), t) \leq 0$) and problems with fixed time, necessary extremum conditions are derived as before by introducing a new variable $x_{n+1}(t)$:

$$\frac{dx_{n+1}(t)}{dt} = 1, \quad x_{n+1}(t_0) = t_0.$$

We conclude with a few words on problems involving the minimization of a functional of the type $\max_{t_0 \leq t \leq t_1} G(x(t), t)$ (the minimax problem), rather than an integral functional. The cone of decrease for this functional was studied in Example 7.5. It is therefore easy to derive results analogous to Theorems 14.1 and 14.2 for the minimax problem. We confine ourselves to a statement of the final result — the analog of Theorem 14.2.

Problem 14.3. Minimize the functional

$$\max_{t_0 \leqslant t < t_1} Gt\,(x\,(t))$$

with respect to $x(t) \in C^{(n)}(t_0, t_1)$, bounded measurable $u(t)$, and t_1, under the constraints

$$\frac{dx(t)}{dt} = \varphi\,(x(t), u(t)), \quad x(t_0) = c,$$

$$x(t_1) = d,$$

$$u(t) \in M \text{ for almost all } t_0 \leqslant t \leqslant t_1.$$

Theorem 14.3. Let $G(x)$ be continuously differentiable with respect to x, $G_x(x) \neq 0$ when $G(x) = \max_{t_0 \leq t \leq t_1} G(x^0(t))$, $\varphi(x, u)$ continuous in x, u and continuously differentiable with respect to x, $M \subset R^r$ closed. Let $x^0(t)$, $u^0(t)$ be a solution of Problem 14.3. Then there exist $\Psi(t)$, a vector $a \in R^n$ and a nonnegative measure $d\mu(t)$ with support on the set

$$R = \{t \in [t_0, t_1]:\ G(x^0(t)) = \max_{t_0 \leqslant t \leqslant t_1} G(x^0(t))\}$$

such that

$$-\psi(t) = -a - \int_t^{t_1} \varphi_x^*(x^0, u^0)\,\psi(\tau)\,d\tau + \int_t^{t_1} G_x(x^0(\tau))\,d\mu(\tau),$$

$$H(x^0(t), \psi(t), u^0(t), t) = 0 \text{ for almost all } t_0 \leqslant t \leqslant t_1,$$

$$H(x^0(t), \psi(t), u(t), t) \leqslant 0 \text{ for almost all } t_0 \leqslant t \leqslant t_1 \text{ and all } u \in M.$$

Here $H(x, \psi, u, t) = (\varphi(x, u), \psi(t))$.

Lecture 15

SUFFICIENT EXTREMUM CONDITIONS

We now return to the general theory of extremal problems, and try to determine whether the necessary conditions for a minimum developed in Lecture 6 are also sufficient. Of course, elementary examples show that in general this is not true. Nevertheless, we shall prove that, under certain additional assumptions, the necessary extremum conditions are also sufficient, in an important class of extremal problems — convex problems. Sufficient conditions for non-convex problems can probably be formulated in terms of the second variation. However, results in this field are as yet quite sporadic, and we shall not dwell on them here.

A convex problem is an extremum problem in which the performance index (i.e., the functional to be minimized) and the constraints are convex. These problems possess an important property: the local and global minima coincide (recall that x_0 is a point of global minimum for $F(x)$ on Q if $F(x_0) \leq F(x)$ for all $x \in Q$; it is a point of local minimum if there is a neighborhood U of x_0 such that $F(x_0) \leq F(x)$ for all $x \in Q \cap U$).

Theorem 15.1. Let $F(x)$ be a convex functional and Q a convex set in a topological linear space E. Then every point of local minimum for $F(x)$ on Q is also a point of global minimum.

Proof. Let x_0 be a local minimum point, so that $F(x_0) \leq F(x)$ for $x \in Q \cap U$, where U is a neighborhood of x_0. Let x_1 be an arbitrary point in Q. Since U is an absorbing set (see Lecture 2), there exists $0 < \lambda < 1$ such that $x_\lambda = x_0 + \lambda(x_1 - x_0) \in U$. Since Q is convex, it follows that $x_\lambda \in Q$, and therefore $F(x_\lambda) \geq F(x_0)$. Since $F(x)$ is convex, we have

$$F(x_\lambda) \leqslant \lambda F(x_0) + (1 - \lambda) F(x_1),$$

and so

$$F(x_1) \geqslant \frac{1}{1-\lambda}(F(x_\lambda) - \lambda F(x_0)) \geqslant F(x_0).$$

But this means that x_0 is a global minimum point.

We can now state sufficient conditions for an extremum.

Theorem 15.2. Let $F(x)$ be a convex continuous functional, $Q_1, \ldots, Q_{n+1}$ convex sets, and assume that there exists a point $\tilde{x}$ such that $\tilde{x} \in \overset{0}{Q}_i$, $i = 1, \ldots, n$, $\tilde{x} \in Q_{n+1}$. Let $x_0 \in Q = \bigcap_{i=1}^{n+1} Q_i$; let K_0 be the cone of decrease of $F(x)$ at x_0, $K_1, \ldots, K_n$ the cones of feasible directions for $Q_1, \ldots, Q_n$, K_{n+1} the tangent cone for Q_{n+1}. Then x_0 is a minimum point for $F(x)$ on Q if and only if there exist $f_i \in K_i^*$, $i = 0, 1, \ldots, n+1$, not all zero, such that

$$f_0 + f_1 + \ldots + f_{n+1} = 0, \tag{15.1}$$

Proof. Necessity. If K_0 is not empty, the result follows from the Dubovitskii-Milyutin theorem (since all the cones $K_0, \ldots, K_n$ are convex, nonempty and open). But if K_0 is empty and x_0 is a minimum point, we can take arbitrary nonzero $f_i \in K_i^*$, $i = 1, \ldots, n+1$, and set $f_0 = -\sum_{i=1}^{n+1} f_i$. Since K_0 is empty, $K_0^* = E'$ and therefore $f_0 \in K_0^*$.

Sufficiency. Let $x_1 \in Q$, $F(x_1) < F(x_0)$. Construct $x_\lambda = \lambda\tilde{x} + (1-\lambda)x_1$, $0 < \lambda < 1$. Since $\tilde{x} \in Q_i$, $x_1 \in Q_i$, $i = 1, \ldots, n+1$, it follows that also $x_\lambda \in Q_i$ (for the sets Q_i are convex), and since $\tilde{x} \in \overset{0}{Q}_i$, $i = 1, \ldots, n$, we have $x_\lambda \in \overset{0}{Q}_i$, $i = 1, \ldots, n$ (Lemma 2.4). Take $\lambda > 0$ so small that $F(x_\lambda) < F(x_0)$ (that this can be done follows from the continuity of $F(x)$, since $F(x_1) < F(x_0)$). Set $h = x_\lambda - x_0$. Then, for $0 < \varepsilon < 1$,

$$F(x_0 + \varepsilon h) \leqslant \varepsilon F(x_\lambda) + (1-\varepsilon) F(x_0)$$

and hence

$$F'(x_0, h) = \lim_{\varepsilon \to +0} \frac{F(x_0 + \varepsilon h) - F(x_0)}{\varepsilon} \leqslant$$

$$\leqslant \lim_{\varepsilon \to +0} \frac{\varepsilon (F(x_\lambda) - F(x_0))}{\varepsilon} = F(x_\lambda) - F(x_0).$$

Therefore (see Theorem 7.4), $h \in K_0$. Furthermore, $h \in K_i$, $i = 1, \ldots, n$, since $x_0 + \varepsilon h \in \overset{0}{Q}_i$, $0 < \varepsilon < 1$, and $h \in K_{n+1}$, since $x_0 + \varepsilon h \in Q_{n+1}$, $0 < \varepsilon < 1$. Thus, $h \in \bigcap_{i=0}^{n+1} K_i$, and the cones K_i, $i = 0, \ldots, n$, are open and convex. Now, by Lemma 5.11, this contradicts condition (15.1). Thus the assumption $F(x_1) < F(x_0)$ must be false, and x_0 is a minimum point.

Let us reformulate Theorem 15.2 for the case in which the sets Q_i, $i = 1, \ldots, n$, are defined by smooth functionals. The problem to be solved is thus

$$\begin{gathered} \min F_0(x), \\ F_i(x) \leqslant 0, \quad i = 1, \ldots, k, \\ F_i(x) = 0, \quad i = k+1, \ldots, n, \quad x \in Q, \end{gathered} \tag{15.2}$$

where $F_i(x)$, $i = 0, \ldots, k$, are convex differentiable functionals, $F_i(x)$, $i = k+1, \ldots, n$ have the form

$$F_i(x) = f_i(x) + \alpha_i, \quad f_i \in E', \ \alpha_i \in R^1, \quad i = k+1, \ldots, n,$$

and Q is a convex set.

T h e o r e m 15.3. Under the above conditions, assume that there exists a point $\tilde{x} \in \overset{0}{Q}$ such that $F_i(\tilde{x}) < 0$, $i = 1, \ldots, k$, $F_i(\tilde{x}) = 0$, $i = k+1, \ldots, n$ (Slater's condition). (If $k = n$ or Q is a polyhedron, the condition $\tilde{x} \in \overset{0}{Q}$ may be replaced by $\tilde{x} \in Q$.) Then $F(x)$ has a minimum at the point x_0 if and only if there exist numbers $\lambda_1, \ldots, \lambda_n$ such that $\lambda_i \geqslant 0$, $\lambda_i F_i(x_0) = 0$, $i = 1, \ldots, k$, and the functional

$$f = F_0'(x_0) + \sum_{i=1}^{n} \lambda_i F_i'(x_0)$$

is a support for Q at the point x_0.

Proof. Necessity follows from Theorem 11.4 and Remark 3 to Theorem 6.1 (since $\check{x} - x_0$ is a feasible direction). We prove sufficiency. Let x_1 be an arbitrary admissible point, i.e., $x_1 \in Q$, $F_i(x_1) \leqslant 0$, $i = 1, \ldots, k$, $F_i(x_1) = 0$, $i = k+1, \ldots, n$. Since the functionals in question are convex (see (7.4)),

$$F_i(x_1) \geqslant F_i(x_0) + (F_i'(x_0), x_1 - x_0), \quad i = 0, \ldots, k,$$

and since the functionals $F_i(x)$, $i = k+1, \ldots, n$, are linear,

$$F_i(x_1) = F_i(x_0) + (F_i'(x_0), x_1 - x_0).$$

Multiply the inequalities for $F_i(x)$, $i = 1, \ldots, k$, by nonnegative numbers λ_i, and the equalities by numbers λ_i, $i = k+1, \ldots, n$, of arbitrary sign; addition of the results gives

$$F_0(x_1) + \sum_{i=1}^{n} \lambda_i F_i(x_1) \geqslant F_0(x_0) + \sum_{i=1}^{n} \lambda_i F_i(x_0) + (f, x_1 - x_0).$$

Since

$$F_i(x_1) \leqslant 0, \quad i = 1, \ldots, k, \quad F_i(x_1) = 0, \quad i = k+1, \ldots, n,$$
$$\lambda_i F_i(x_0) = 0, \quad i = 1, \ldots, k,$$

it now follows that

$$F_0(x_1) \geqslant F_0(x_0) + (f, x_1 - x_0).$$

But f is a support to Q at x_0, and since $x_1 \in Q$ we have $(f, x_1 - x_0) \geqslant 0$. Thus, finally, we get $F_0(x_1) \geqslant F_0(x_0)$, so that x_0 is a point of (absolute) minimum.

Remark. The proof of sufficiency made no use of the assumption that $\tilde{x}$ exists.

We now formulate the extremum conditions for the same problem in a different form, in terms of the saddle point of the Lagrange function. The Lagrange function is defined as

$$L(x, y) = F_0(x) + \sum_{i=1}^{n} y_i F_i(x), \tag{15.3}$$

where $x \in E$, $y \in R^n$. A saddle point of a function $L(x, y)$ of two variables in a domain $x \in Q_1$, $y \in Q_2$ is a pair $x_0 \in Q_1$, $y_0 \in Q_2$ such that x_0 is a minimum point for $L(x, y_0)$ on Q_1 and y_0 is a maximum point for $L(x_0, y)$ on Q_2, i.e.,

$$L(x_0, y) \leqslant L(x_0, y_0) \leqslant L(x, y_0)$$

for all $x \in Q_1$, $y \in Q_2$.

Theorem 15.4. Under the assumptions of Theorem 15.3, the following condition is necessary and sufficient for an extremum in Problem (15.2) at the point x_0: There exists $y_0 = (\lambda_1, \ldots, \lambda_n)$ such that x_0, y_0 is a saddle point of the Lagrange function $L(x, y)$ in the domain

$$x \in Q, \quad y \in R = \{y = (y_1, \ldots, y_n) \colon y_i \geqslant 0, \quad i = 1, \ldots, k\}.$$

Proof. Let x_0 be a solution of (15.2). Then, by Theorem 15.3, there exist

$$\lambda_1, \ldots, \lambda_n, \quad \lambda_i \geqslant 0, \quad i = 1, \ldots, k, \quad \lambda_i F_i(x_0) = 0, \quad i = 1, \ldots, k,$$

such that

$$f = F_0'(x_0) + \sum_{i=1}^{n} \lambda_i F_i'(x_0)$$

is a support to Q at x_0. We claim that then x_0, y_0, where $y_0 = (\lambda_1, \ldots, \lambda_n)$, is a saddle point. Indeed, for any $y \in R^n$,

$$L(x_0, y) = F_0(x_0) + \sum_{i=1}^{n} y_i F_i(x_0) = F_0(x_0) + \sum_{i=1}^{k} y_i F_i(x_0) \leqslant$$
$$\leqslant F_0(x_0) = L(x_0, \bar{y}_0),$$

since

$$y_i \geqslant 0,\ i=1,\ldots,k,\ F_i(x_0) \leqslant 0,\ i=1,\ldots,k,\ y_i F_i(x_0) = 0,\ i=k+1,\ldots,n.$$

Similarly, for any $x \in Q$ we get, using the definition of a supporting functional and the property of convex functionals,

$$L(x, y_0) = F_0(x) + \sum_{i=1}^{n} \lambda_i F_i(x) \geqslant F_0(x_0) + \sum_{i=1}^{n} \lambda_i F_i(x_0) +$$
$$+ (f, x - x_0) \geqslant L(x_0, y_0).$$

Conversely, let $x_0, y_0 = (\lambda_1, \ldots, \lambda_n)$ be a saddle point. Then, for $y \in R^n$,

$$0 \leqslant L(x_0, x_0) - L(x_0, y) = \sum_{i=1}^{n} (\lambda_i - y_i) F_i(x_0).$$

This inequality can hold for arbitrary y_i, $i = k+1, \ldots, n$, only if $F_i(x_0) = 0$, $i = k+1, \ldots, n$, and for arbitrary $y_i \geqslant 0$, $i = 1, \ldots, k$, only if $F_i(x_0) \leqslant 0$, $i = 1, \ldots, k$; moreover, either λ_i or $F_i(x_0)$ must vanish, i.e., $\lambda_i F_i(x_0) = 0$, $i = 1, \ldots, k$. Thus x_0 satisfies all the constraints of Problem (15.2). Now, since $L(x, y_0)$ assumes a minimum on Q at the point x_0, the necessary condition for a minimum must hold at x_0, i.e., $L_x(x_0, y_0)$ is a support to Q at the point x_0. But

$$L_x(x_0, y_0) = F_0'(x_0) + \sum_{i=1}^{n} \lambda_i F_i'(x_0).$$

Hence the sufficient condition for an extremum in the original problem is satisfied (Theorem 15.3), and so x_0 is a solution.

R e m a r k. It can be shown that Theorem 15.4 remains valid for nondifferentiable functionals.

Lecture 16

SUFFICIENT EXTREMUM CONDITIONS.

EXAMPLES

We now apply the results of the preceding lecture to various problems.

1. Linear programming problem

$$\min (c, x), \quad Ax \leqslant b, \quad x \geqslant 0. \tag{16.1}$$

Here

$$x \in R^m, \quad c \in R^m, \quad (c, x) = \sum_{i=1}^{m} c_i x_i, \quad b \in R^n,$$

A is an $n \times m$ matrix, and $Ax \leqslant b$, $x \geqslant 0$ means that

$$(Ax)_i \leqslant b_i, \quad i = 1, \ldots, n, \quad x_i \geqslant 0, \quad i = 1, \ldots, m.$$

In Lecture 11, we found necessary conditions for a minimum in this problem: If x^0 is a solution, there exists $y^0 \in R^n$ such that $y^0 \geqslant 0$ and

$$A^* y^0 \geqslant c; \quad y_i^0 (Ax^0 - b)_i = 0, \quad i = 1, \ldots, n.$$

It now follows from Theorem 15.3 that this condition is sufficient. Moreover, Theorem 15.4 shows that that pair x^0, y^0 is a saddle point for the Lagrange function

$$L(x, y) = (c, x) + (Ax - b, y)$$

in the domain $x \geqslant 0$, $y \geqslant 0$.

Together with problem (16.1), we consider another problem, in the space R^n:

$$\max (b, y), \quad A^*y \geqslant c, \quad y \geqslant 0. \tag{16.2}$$

Either of problems (16.1) and (16.2) (the first is known as the primal problem and the second as the dual problem) can be derived from the other by the following formal rules. a) Replace minimization by maximization and vice versa. b) Any constraint vector becomes a functional, a functional vector becomes a constraint. c) Replace the matrix by its transpose. d) Reverse the direction of the inequalities. e) The nonnegativity condition for the variables is retained. These rules are symmetric, so that (16.2) is the dual of (16.1) and (16.1) the dual of (16.2). One therefore speaks simply of a pair of dual problems.

Theorem 16.1 (Duality principle). Dual problems are either both solvable or both unsolvable. When a solution exists, the extremal values of the functionals are equal (i.e., $(c, x^0) = (b, y^0)$), and any pair of solutions x^0, y^0 satisfies the so-called complementarity conditions:

$$x_i^0(A^*y^0 - c)_i = 0, \quad i = 1, \dots, m, \quad (Ax^0 - b)_i\, y_i^0 = 0, \quad i = 1, \dots, n.$$

Proof. Let one of the problems (the first, say) have a solution x^0. Then there exists $y^0 \geqslant 0$ such that the pair x^0, y^0 is a saddle point of the Lagrange function $L(x, y) = (c, x) + (Ax - b, y)$ in the domain $x \geqslant 0$, $y \geqslant 0$. Since

$$L(x, y) = (-b, y) + (A^*y - c, x) = -\bar{L}(y, x),$$

where $\bar{L}(y, x)$ is the Lagrange function for the second problem (in which the independent variables are denoted by y and the Lagrange multipliers by x), it follows that the pair x^0, y^0 is also a saddle point for the second problem, and so y^0 is a solution. Then $L(x^0, y^0) = (c, x^0)$, and $\bar{L}(y^0, x^0) = -(b, y^0)$; therefore $(c, x^0) = (b, y^0)$, and it follows from Theorem 15.3 that

$$(Ax^0 - b)_i y_i^0 = 0, \quad i = 1, \dots, n, \quad (A^* y^0 - c)_i x_i^0 = 0 \quad i = 1, \dots, m.$$

2. Optimal control problem for a linear system

$$\min \int_0^T F(x(t), u(t), t)\, dt,$$

$$\frac{dx(t)}{dt} = A(t)x(t) + B(t)u(t),$$

$$x(0) = c, \quad x(T) = d,$$

$$u(t) \in M \text{ for almost all } 0 \leqslant t \leqslant T,$$

$$G(x(t), t) \leqslant 0 \text{ for all } 0 \leqslant t \leqslant T.$$

We assume here that $F(x, u, t)$ is convex and continuously differentiable with respect to x, u and continuous in t, $A(t)$, $B(t)$ are matrices which depend continuously on t, M is a closed convex set in R^r, $M^0 \neq \emptyset$, $G(x, t)$ is convex and continuously differentiable with respect to x and continuous in t, $G_x(x, t) \neq 0$ when $G(x, t) = 0$, $G(c, 0) < 0$, $G(d, T) < 0$. Assume moreover that the nondegeneracy condition holds (see Lecture 9): $B^*(t)\Psi(t) \not\equiv 0$ for every nonzero solution $\Psi(t)$ of the system

$$\frac{d\psi(t)}{dt} = -A^*(t)\psi(t).$$

Finally, assume that there exists $\tilde{u}(t)$ such that $\tilde{u}(t) \in M^0$ for almost all $0 \leq t \leq T$, and the corresponding function $\tilde{x}(t)$ is such that $\tilde{x}(T) = d$, $G(\tilde{x}(t), t) < 0$ for $0 \leq t \leq T$.

Theorem 16.2. Under the above assumptions, the local maximum principle (Theorem 14.1) is a sufficient condition for an extremum.

In fact, an examination of the proof of Theorem 14.1 shows that this statement is simply an application of the result of Theorem 15.2 to our problem.

There are various generalizations of Theorem 16.2 (relaxation of the nondegeneracy condition, other types of functionals and boundary conditions, etc.), but we shall not discuss them here.

SUGGESTIONS FOR FURTHER READING

Lecture 1. There are many excellent texts on the calculus of variations, e.g., Akhiezer [1], Bliss [1], Gel'fand and Fomin [1], Gyunter [1], Lavrent'ev and Lyusternik [1], [2], and also Tslaf [1].

Extremum conditions for a smooth functional in the presence of smooth constraints were first obtained in general form (in Banach spaces) by Lyusternik [1] in 1934 (see also Goldstine [1]).

The methods of approximation theory are described, e.g., in Akhiezer [2].

The maximum principle for the optimal control problem was stated as a conjecture in the paper of Boltyanskii, Gamkrelidze and Pontryagin [1] and first proved by Boltyanskii [1] in 1958. A detailed exposition of the theory developed by these authors may be found in Pontryagin, Boltyanskii, Gamkrelidze and Mishchenko [1] or (in simplified form) in Boltyanskii [2].

Problems with nonsmooth constraints, which arise in the field of economics, were first considered by Kantorovich [1], who also gave the first formulation of necessary extremum conditions in the general problem with nonsmooth constraints (Kantorovich [2]). Extremum conditions were obtained for the linear programming problem by Dantzig [1], and for the nonlinear programming problem by John [1] and Kuhn and Tucker [1]. Numerous texts are now available [in Russian] on linear and nonlinear programming: Gass [1], Dantzig [2], Dennis [1], Zoutendijk [1], Zukhovitskii and Avdeeva [1], Karlin [1], Kuhn and Tucker [2], Künzi and Krelle [1], Hadley [1], Arrow, Hurwicz and Uzawa [1], Yudin and Gol'shtein [1], [2].

The most general conditions for an extremum in general problems with constraints were first obtained by Dubovitskii and Milyutin [1] in 1962. A detailed account of their theory is given in their paper Dubovitskii and Milyutin [2], and also in Milyutin's doctoral dissertation [1]. The elements of the theory of the second variation for problems with constraints are presented in Dubovitskii and Milyutin [4].

Lectures 2, 3, 4. Material from the theory of topological linear spaces may be found in Bourbaki [1], Dunford and Schwartz [1], Kantorovich and Akilov [1], Robertson and Robertson [1], Arrow, Hurwicz and Uzawa [1]. For the theory of Banach spaces the reader may consult Day [1], Kolmogorov and Fomin [1], Lyusternik and Sobolev [1], Riesz and Sz.-Nagy [1]. In particular, the proofs of Theorems 3.1, 3.2, 3.3, 3.4, 4.2 and 4.3 may be found in Dunford and Schwartz [1].

Lecture 5. Theorem 5.1 is due to Krein [1]. Another proof is given in Bourbaki [1]. The important Theorem 5.2 was first proved by Dubovitskii and Milyutin and published without proof in their paper [1]. A remarkably simple proof of the theorem is given by Milyutin [1].

Lecture 6. The first general statement of extremum problems in function spaces is apparently due to Kantorovich [2]. In the same paper, he obtained necessary extremum conditions for such problems in terms of linear functionals. In their final form, the conditions were stated by Dubovitskii and Milyutin in [1], [2] without proof (for the proofs, see Milyutin [1]). The essential difference between these results and those of Kantorovich [2] is in our opinion as follows. First, Dubovitskii and Milyutin consider a constraint defined by the intersection of several sets, and the extremum conditions involve linear functionals each related to one of these sets. It thus becomes possible to carry out separate analysis of each constraint (rather than of their intersection, which is far more difficult). Second, in the Dubovitskii-Milyutin scheme one can consider nonconvex constraints, and also equality constraints. Third, Dubovitskii and Milyutin investigate many constraints in concrete problems. With this information in hand, it is no more difficult to derive extremum conditions in problems that involve these constraints (in arbitrary combinations) than, say, to differentiate some complicated function using the rule for differentiation of a composite function. Finally, Dubovitskii and Milyutin show that, if the space in which the extremal problem is treated is properly chosen, many extremum conditions which previously were of global nature (such as the Pontryagin maximum principle) appear in the guise of local extremum conditions, and they may therefore be attacked using the general technique.

The presentation in Lecture 6 is somewhat different from the scheme of Dubovitskii and Milyutin [1], [2]: the definitions of feasible and tangent directions, directions of decrease, are slightly different; linearly-convex functionals are not used; the spaces considered are the more general topological linear spaces, rather than Banach spaces; and so on.

A considerable number of publications have appeared recently which discuss extremum conditions in general problems with constraints: Pshenichnyi [1], Gamkrelidze and Kharatashvili [1], [2], Dem'yanov and Rubinov [1], Neustadt [1], [2], [3], Halkin and Neustadt [1], Halkin [1], [2], [4], Hestenes [1]. In principle, most of this work resembles the Dubovitskii-Milyutin scheme. A specific feature of some of these papers is that they employ separability of sets not in the initial space but in a finite-dimensional image space (it is assumed that the constraints are given by a finite number of not necessarily smooth functionals).

Another approach to extremal problems is closely bound up with the concept of duality. In this connection, see Gol'shtein [1], [2], Ioffe and Tikhomirov [1], and Rubinshtein [1], [2].

Lectures 7, 8. The examples presented here were considered by Dubovitskii and Milyutin [2].

Lecture 9. Theorem 9.1 is due to Lyusternik, and a proof may be found in Lyusternik and Sobolev [1]. Use of this theorem provides the simplest method for determining tangent subspaces (Dubovitskii and Milyutin employ a different method to determine the tangent subspaces in Examples 9.2 and 9.3).

Differentiation of nonlinear operators is examined in detail in the books of Vainberg [1], Kantorovich and Akilov [1], Lyusternik and Sobolev [1].

L e c t u r e 10. The finite-dimensional case of Theorem 10.4 has been known, in a variety of formulations, for some time (e.g., Minkowski and Farkas). Since this theorem is fundamental for the derivation of extremum conditions in problems of mathematical programming, a proof of some version may be found in practically any book on linear or nonlinear programming. The infinite-dimensional case of the theorem is proved, by a method different from ours, in Arrow, Hurwicz and Uzawa [1].

L e c t u r e 11. The rule of Lagrange multipliers was generalized to Banach spaces by Lyusternik [1] in 1934, and later by Goldstine [1]. Linear programming problems were first studied by Kantorovich [1] in 1939, but the rigorous statement of the extremum conditions is due to Dantzig [1]. Theorem 11.2 in the finite-dimensional case was first proved, apparently, by John [1] in 1948 (and not by Kuhn and Tucker [1], as is often thought). However, a similar result follows from the earlier work of Cox [1].

L e c t u r e s 12, 13. The maximum principle was proposed as a conjecture in 1956, by Boltyanskii, Gamkrelidze and Pontryagin [1], and proved by Boltyanskii [1] in 1958. However, Hestenes [1] claims to have proved the maximum principle as early as 1950, in a Technical Report published by RAND Corporation (RM—100). The method of proof used in Lectures 12 and 13 is due to Dubovitskii and Milyutin [1], [2]. Several other proofs of the maximum principle are known. We mention only two of the shortest ones. One, due to Rozonoer [1], is based on estimates for increments to the functional, of the type

$$F(u+\bar{u})-F(u)=\int (H(x,\, u+u,\, \psi,\, t)-H(x,\, u,\, \psi,\, t))\, dt+o(\|\bar{u}\|);$$

the other, which uses Lyapunov's theorem on vector-valued measures, is due to Halkin [3].

L e c t u r e 14. Extremum conditions in optimal control problems with constraints were obtained by Gamkrelidze [1]. His exposition, however, requires some additional assumptions (the so-called regularity condition). The maximum principle as stated in Theorem 13.2 was stated and proved by Dubovitskii and Milyutin [1], [2]. It can be shown that Gamkrelidze's result follows from Theorem 13.2. At present, many other (apparently equivalent) formulations of the extremum conditions for this problem are known.

L e c t u r e 15. A similar form of the sufficient conditions for an extremum in the general convex problem was obtained by Pshenichnyi [1], using another method. The saddle-point theorem for convex programming problems in the finite-dimensional case was first proved by Kuhn and Tucker [1].

L e c t u r e 16. The duality theorem for linear programming problems was first proved by Dantzig [1]. More far-reaching generalizations have been obtained by Gol'shtein [1], [2], Ioffe and Tikhomirov [1], and Rubinshtein [1], [2].

One generalization of Theorem 16.2 is due to Dubovitskii and Milyutin [3], who prove the sufficiency of the maximum principle in time-optimal problems. Rozonoer [2] has proved the sufficiency of the maximum principle for systems which are linear only in x (but not in u). Polyak [1] has proved that the local maximum principle furnishes sufficient conditions for nonlinear systems, but this is done under quite restrictive assumptions on the integrand. Sufficient conditions for optimal control problems, based on the ideas of dynamic programming (Bellman [1]), have been obtained by Boltyanskii [3] and Krotov [1]. It was Girsanov [1] who pointed out that Krotov's approach is related to dynamic programming.

REFERENCES

[Books and papers in Russian are indicated by an asterisk.]

Akhiezer, N.I.

*1. Lectures on the Calculus of Variations. Moscow, Gostekhizdat, 1955.

*2. Lectures on Approximation Theory. Moscow, Nauka, 1965.

Arrow, K.J., Hurwicz, L., Uzawa, H. (eds.)

1. Studies in Linear and Nonlinear Programming. Stanford University Press, 1958.

Bellman, R.E.

1. Dynamic Programming. Princeton University Press, 1957.

Bliss, G.A.

1. Lectures on the Calculus of Variations. University of Chicago Press, 1946.

Boltyanskii, V.G.

*1. The maximum principle in the theory of optimal processes. Dokl. Akad. Nauk SSSR 119, No. 6 (1958), 1070–1073.

*2. Mathematical Methods of Optimal Control. Moscow, Nauka, 1966.

*3. Sufficient optimum conditions and justification of the method of dynamic programming. Izv. Akad. Nauk SSSR Ser. Mat. 28, No. 3 (1964), 481–514.

Boltyanskii, V.G., Gamkrelidze, R.V., Pontryagin, L.S.

*1. Toward a theory of optimal processes. Dokl. Akad. Nauk SSSR 110, No. 1 (1956), 7–10.

Bourbaki, N.

1. Eléments de Mathématique, Première Partie, Livre V: Espaces Vectoriels Topologiques. Paris, Hermann, 1953.

Cox, M. J.

1. On necessary conditions for relative minimum. Amer. J. Math. 66, No. 2 (1944), 170–198.

Dantzig, G. B.

1. Programming of interdependent activities, II. Econometrica 17, No. 3 (1949), 200–201.

2. Linear Programming and Extensions. Princeton University Press, 1963.

Day, M. M.

1. Normed Linear Spaces. Berlin, Springer-Verlag, 1958.

Dem'yanov, V. F., Rubinov, A. M.

*1. Approximate Methods for Solution of Extremal Problems. Leningrad, Leningrad State University, 1968.

Dennis, J. B.

1. Mathematical Programming and Electrical Networks. Cambridge, Mass., Technology Press of the Massachusetts Institute of Technology, 1959.

Dubovitskii, A. Ya., Milyutin, A. A.

*1. The extremum problem in the presence of constraints. Dokl. Akad. Nauk SSSR 149, No. 4 (1963), 759–762.

*2. Extremum problems in the presence of constraints. Zh. Vychisl. Mat. i Mat. Fiz. 5, No. 3 (1965), 395–453.

*3. Some optimum problems for linear systems. Avtomat. i Telemekh. 24, No. 12 (1963), 1616–1625.

*4. Second variations in extremum problems with constraints. Dokl. Akad. Nauk SSSR 160, No. 1 (1965), 18–21.

Dunford, N., Schwartz, J.

1. Linear Operators, Part I. New York, Interscience Publishers, 1958.

Gamkrelidze, R. V.

*1. Time-optimal processes with constrained phase coordinates. Dokl. Akad. Nauk SSSR 125, No. 3 (1959), 475—478.

Gamkrelidze, R. V., Kharatashvili, G. L.

*1. Theory of the first variation in extremal problems. Soobshch. Akad. Nauk Gruz. SSR 46, No. 1 (1967), 27—31.

2. Extremal problems in linear topological space, I. Math. Syst. Theory 1, No. 3 (1967), 229—256.

Gass, S.

1. Linear Programming Methods and Applications. New York, McGraw-Hill, 1958.

Gel'fand, I. M., Fomin, S. V.

*1. Calculus of Variations. Moscow, Fizmatgiz, 1961.

Girsanov, I. V.

*1. Some relations between the functions of Bellman and Krotov for problems of dynamic programming. Vestnik Moskov. Univ. Ser. I Mat. Mekh., No. 2 (1968), 56—59.

Goldstine, H. H.

1. Condition for a minimum in abstract space. Illinois J. Math. 2, No. 1 (1958).

Gol'shtein, E. G.

*1. Dual problems of convex and convex-fractional programming. Dokl. Akad. Nauk SSSR 172, No. 5 (1967), 1007—1010.

*2. Dual problems of convex and convex-fractional programming in function spaces. In: Studies in Mathematical Programming, 10—108. Moscow, Nauka, 1968.

Gyunter, N. M.

*1. Course in the Calculus of Variations. Moscow-Leningrad, Gostekhizdat, 1941.

Hadley, G.

1. Nonlinear and Dynamic Programming. London, Addison-Wesley, 1964.

Halkin, H.

1. On the necessary condition for optimal control of nonlinear systems. J. Analyse Math. 12, No. 1 (1964), 1—82.

2. An abstract framework for the theory of process optimization. Bull. Amer. Math. Soc. 72, No. 4 (1966), 677—678.

3. Liapunov's theorem on the range of a vector measure and Pontryagin's maximum principle. Arch. Rational Mech. Anal. 10, No. 4 (1962), 296—304.

4. Nonlinear nonconvex programming in an infinite dimensional space. In: Mathematical Theory of Control, ed. A. V. Balakrishnan and L. W. Neustadt, 10—25. New York, Academic Press, 1967.

Halkin, H., Neustadt, L. W.

1. General necessary conditions for optimization problems. Proc. Nat. Acad. Sci. U. S. A. 54, No. 4 (1966), 1066—1071.

Hestenes, M. R.

1. Calculus of Variations and Optimal Control Theory. New York, John Wiley and Sons, 1966.

Ioffe, A. D., Tikhomirov, V. M.

*1. Duality of convex functions and extremal problems. Uspekhi Mat. Nauk 23, No. 6 (1968), 51—116.

John, F.

1. Extremum problems with inequalities as side conditions. In: Studies and

Essays (Courant Anniversary Volume), ed. K. O. Friedrichs et al., 187–204. New York, 1948.

Kantorovich, L. V.

*1. Mathematical Methods in Organization and Planning of Production. Leningrad, Leningrad State University, 1939.

*2. On an effective method for solution of certain extremal problems. Dokl. Akad. Nauk SSSR 28, No. 3 (1940), 212–215.

Kantorovich, L. V., Akilov, G. P.

*1. Functional Analysis in Normed Spaces. Moscow, Fizmatgiz, 1959.

Karlin, S.

1. Mathematical Methods and Theory in Games, Programming and Economics. London, Addison-Wesley, 1959.

Kolmogorov, A. N., Fomin, S. V.

*1. Elements of the Theory of Functions and Functional Analysis. Moscow, Nauka, 1968.

Krein, M. G.

*1. On positive functionals in linear normed spaces. In: Some Problems in the Theory of Moments (eds. N. I. Akhiezer and M. G. Krein). Khar'kov, State United Publishing House of Science and Technology (GONTI), 1938.

Krotov, V. F.

*1. Methods for solution of variational problems based on sufficient conditions for an absolute minimum, I, II, III. Avtomat. i Telemekh. 23, No. 12 (1962), 1571–1583; 24, No. 5 (1963), 581–598; 25, No. 7 (1964), 1037–1046.

Kuhn, H. W., Tucker, A. W.

1. Nonlinear programming. Proceedings of the Second Berkeley Symposium on Mathematical Statistics and Probability, 481–492. Berkeley, 1950.

2. (eds.) Linear Inequalities and Related Systems. Princeton University Press, 1956.

Künzi, H. P., Krelle, W.

1. Nichtlineare Programmierung. Berlin, Springer-Verlag, 1962.

Lavrent'ev, M. A., Lyusternik, L. A.

*1. Elements of the Calculus of Variations, Vol. 1, Parts 1 and 2. Moscow-Leningrad, Department of Scientific and Technical Information (ONTI), 1935.

*2. Course in the Calculus of Variations. Moscow-Leningrad, Gostekhizdat, 1950.

Lyusternik, L. A.

*1. On extremum conditions for functionals. Mat. Sb. 41, No. 3 (1934), 390—401.

Lyusternik, L. A., Sobolev, V. I.

*1. Elements of Functional Analysis. Moscow, Nauka, 1965.

Milyutin, A. A.

*1. Extremum Problems in the Presence of Constraints (Doctoral Dissertation). Moscow, Institut Prikladnoi Matematiki Akad. Nauk SSSR, 1966.

Natanson, I. P.

*1. Theory of Functions of a Real Variable. Moscow-Leningrad, State Publishing House of Technical and Theoretical Literature (GITTL), 1950.

Neustadt, L. W.

*1. General theory of variational problems with applications to optimal control. Dokl. Akad. Nauk SSSR 171, No. 1 (1966), 48—50.

2. An abstract variational theory with applications to a broad class of optimization problems, I. SIAM J. Control 4 (1966), 505—527.

3. An abstract variational theory with applications to a broad class of optimi-

zation problems, II. SIAM J. Control 5, No. 1 (1967), 90—137.

Polyak, B. T.

*1. On the theory of nonlinear optimal control problems. Vestnik Moskov. Univ. Ser. I Mat. Mekh., No. 2 (1968), 30—40.

Pontryagin, L. S., Boltyanskii, V. G., Gamkrelidze, R. V., Mishchenko, E. F.

*1. Mathematical Theory of Optimal Processes. Moscow, Fizmatgiz, 1961.

Pshenichnyi, B. N.

*1. Convex programming in a normed space. Kibernetika, No. 5 (1965), 46—54.

Riesz, F., Sz.-Nagy

1. Leçons d'Analyse Fonctionelle (2nd edition). Budapest, Akademiai Kiado, 1953.

Robertson, A. P., Robertson, W.

1. Topological Vector Spaces. Cambridge University Press, 1964.

Rozonoer, L. I.

*1. Pontryagin's maximum principle in the theory of optimal systems, I, II, III. Avtomat. i Telemekh. 20, No. 10 (1959), 1320—1334; 20, No. 11 (1959), 1441—1458; 20, No. 12 (1959), 1561—1578.

Rubinshtein, G. Sh.

*1. Dual extremal problems. Dokl. Akad. Nauk SSSR 152, No. 2 (1963), 288—291.

*2. Some examples of dual extremal problems. In: Mathematical Programming, 9—39. Moscow, Nauka, 1966.

Tslaf, L. Ya.

*1. Calculus of Variations and Integral Equations. Moscow, Nauka, 1966.

Vainberg, M. M.

*1. Variational Methods for the Study of Nonlinear Operators. Moscow, Gostekhizdat, 1956.

Yudin, D. B., Gol'shtein, E. G.

*1. Problems and Methods of Linear Programming. Moscow, Sovetskoe Radio, 1964.

*2. Linear Programming. Moscow, Fizmatgiz, 1963.

Zoutendijk, G.

1. Methods of Feasible Directions: A Study in Linear and Nonlinear Programming. Amsterdam, Elsevier Publishing Company, 1960.

Zukhovitskii, S. I., Avdeeva, L. I.

*1. Linear and Convex Programming. Moscow, Nauka, 1967.